Mapping the Deep

MAPPING
THE
DEEP

Innovation, Exploration,
and the Dive of a Lifetime

DAWN J. WRIGHT

with Esri Press

Esri Press
REDLANDS | CALIFORNIA

Esri Press, 380 New York Street, Redlands, California 92373-8100
Copyright © 2024 Esri
All rights reserved.
Printed in the United States of America.

ISBN: 9781589487888
Library of Congress Control Number: 2024932485

Contents

See this book come alive at **mappingthedeep.com.**

Foreword

first met Dawn Wright in 2011, at a meeting of the Science Advisory Board of the National Oceanic and Atmospheric Administration (NOAA), where I was then deputy administrator and chief scientist. In my enthusiasm at encountering a fellow ocean geologist, I gave Dawn a friendly but perhaps overly vigorous punch in the arm—for which, fortunately, she forgave me. We've kept up with each other ever since and, as friends and colleagues, our paths have often crossed.

Dawn and I are both explorers and oceanographers, driven by an intense curiosity about this planet, how it works, and how to put our scientific understanding of the Earth to the best possible use. In addition to sharing an academic discipline and an exceptional experience (we're among the handful of humans who've dived to Challenger Deep, the deepest spot on the planet), I believe Dawn and I share something else: a sense of the interconnectedness of life on this planet, and the urgency of preserving it.

As an astronaut on three space shuttle missions, and the first US woman to walk in space, I was privileged to see our "blue marble" from a unique perspective. It soon became clear to me that all of us on Earth—all biological species—are inextricably linked. This planet is our life support system. Just as an astronaut needs an intimate understanding of their spacesuit to survive a spacewalk, so do we all depend for survival on understanding deeply how our planet works.

Dawn is one of the scientists leading the way in this quest. It's not just her achievements as an oceanographer, an educator, and a trailblazer that are noteworthy, it's her life story as well. Hers is an inspiring story of race and identity, with lessons for all of us on diversity, otherness, and the courage it takes to be the only or the "other" in a place. It's also a tale of succeeding without mentors or role models who look like you, and of holding true to yourself and your goals, no matter what.

The story of Dawn's adventurous life and her historic dive to Challenger Deep makes for a terrific read. I'm sure you will find this book as inspiring and enlightening as I do, an inside account of science and exploration at the edge, expanding our human horizons.

Dr. Kathryn D. Sullivan
NOAA Administrator 2013–2017
NASA Astronaut (retired)
March 1, 2024

ZONES OF THE DEEP

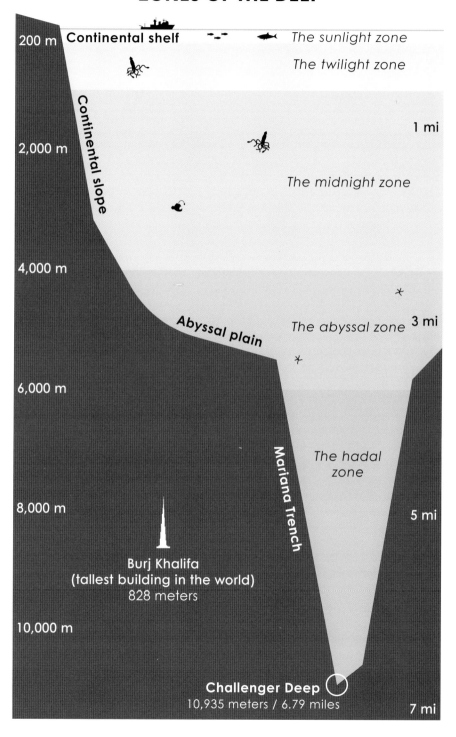

200 m — Continental shelf — The sunlight zone

The twilight zone

Continental slope

1 mi

2,000 m

The midnight zone

4,000 m

Abyssal plain — The abyssal zone — 3 mi

6,000 m

Mariana Trench

The hadal zone

8,000 m — 5 mi

Burj Khalifa
(tallest building in the world)
828 meters

10,000 m

Challenger Deep
10,935 meters / 6.79 miles

7 mi

Chapter 1

THE DIVE

It's a bright summer morning on the seemingly infinite expanse of the western Pacific Ocean, just over 200 miles southeast of the island of Guam. From the deck of the DSSV *Pressure Drop*, nothing but lightly rolling waters can be seen in any direction—that is, except for a spot off the stern of the ship, where you might just make out a curvilinear white shape slipping beneath the surface.

Although, at first glance, it might look like something out of a mid-20th-century science fiction movie, this object is no anachronism. It's one of the most advanced underwater vessels ever devised, the DSV *Limiting Factor*. And this patch of sea, otherwise indistinguishable amid the vastness of the Pacific, is special, too: it's directly above—almost 11 kilometers (7 miles) above—the deepest point in all of Earth's oceans, a notch in the Mariana Trench known as Challenger Deep.

The Mariana Trench is a crescent-shaped chasm in the western Pacific Ocean, spanning more than 2,542 kilometers (1,580 miles); within it lies an even deeper groove, which owes its odd name to the British Royal Navy survey ship HMS *Challenger*, whose scientists first sounded its depths in 1875. Challenger Deep measures about 11 kilometers (7 miles) long by 1.6 kilometers (1 mile) wide and 10,935 meters (35,876 feet) deep. How deep is that, in human terms? It's hard to conceive, but, as an approximation, picture six Grand Canyons stacked on top of each other. Another way to think about it: it's much deeper than Mount Everest is tall.

The *Limiting Factor* is the first and only US-owned vehicle to make multiple descents to Challenger Deep with human beings aboard. (The only other sub that can currently accomplish this is *Fendouzhe*, or "Striver," operated by the Chinese

Dawn Wright prepares to enter the *Limiting Factor* submersible.

government.) Before its initial visit to Challenger Deep in 2019, only two other human-piloted expeditions had ever made it to such depths. But on this fine day a little more than three years later, the *Limiting Factor* is embarking on its 19th such dive. It's an indication of how far deep-sea exploration has come in such a short time that a trek to what was once the most difficult-to-reach place on the planet is now starting to feel almost routine, like catching a flight.

And yet there is nothing ordinary about this dive, especially not for the two people in blue jumpsuits who are sealed inside the submersible's Smart Car–sized cockpit. One of them, Victor Vescovo, is responsible for maneuvering the vessel on its way to and from the bottom of the ocean. It was his quest, launched in 2018, to personally visit the deepest point in every ocean that led to the formation of Caladan Oceanic—a science and technology firm dedicated to increasing humanity's understanding of the deep. (Strictly speaking, our planet has only one, connected ocean, but it's traditionally been divided into five major regions:

the Pacific, Atlantic, Indian, Arctic, and Southern Oceans.) He's piloted the *Limiting Factor* on 15 of its trips to Challenger Deep, but there's a chance this may be the last time he performs that feat. Having achieved his goal—and so much more—he's on the verge of selling the undersea exploration system that he envisioned and piloted through its many record-setting expeditions.

Victor's companion on this descent is the oceanographer Dawn Wright, aka Deepsea Dawn. Now the chief scientist of Esri®, the world's leading geographic information system (GIS) software firm, Dawn has dedicated her life to learning about the ocean—and working to ensure that knowledge of the deep is more widely shared. For Dawn, experiencing Challenger Deep firsthand will be the opportunity of a lifetime.

+0 HOURS, 16 MINUTES | −931 METERS

Darkness comes quickly beneath the ocean's surface. Sunlight penetrates only the uppermost 400 meters (1,312 feet) of water, a threshold the *Limiting Factor* blows through within minutes of sinking under the waves. The next 600 meters (1,968 feet) are fittingly called the twilight zone. This region is believed to be home to more marine life than the rest of the ocean—and much of it provides its own radiance to make up for the lack of natural light. As Dawn says, "It's exhilarating, and it never gets old, the fact that you're descending through the lit zone of the ocean, which is beautiful aqua blue, then things slowly turn to gray and then pitch black."

As the *Limiting Factor* nears the lower extent of the twilight zone, a glow appears through Victor's porthole, off to the sub's left. Believing the source to be some kind of bioluminescent life-form, probably jellyfish or siphonophores (wormlike organisms), Victor flashes the sub's lights. Much to the delight of their human visitors, the creatures respond in kind. This conversation of sorts between nature and machine serves as a fleeting example of the magic of the ocean—a magic that Dawn has felt deeply since her childhood in Hawaii.

Dawn, now 62, moved to Hawaii from the East Coast at the age of six, when her mother accepted a teaching position there. She recalls spending much of her time at the beach, swimming and exploring: "That is part of the culture of Hawaii," she says, "to enjoy but also to hold the ocean as sacred, as life giving. It's a natural part of everyday life there." By the age of eight, inspired by watching Jacques Cousteau on TV, Dawn had decided to become an oceanographer. And yet today,

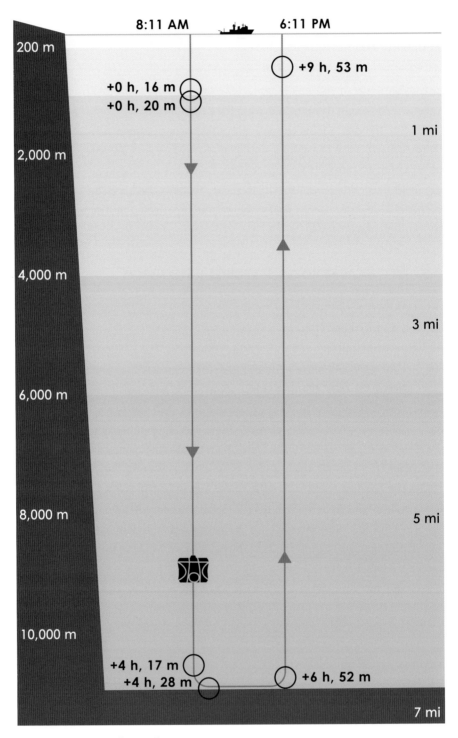

Dawn and Victor's dive path.

Dawn (*left*) and Victor Vescovo ready to board the *Limiting Factor*.

at one of the high points of her career, she's feeling a little wistful, because her mother, who died just months ago, isn't there to witness it.

> " *The ocean has always just been such a natural part of my life ... it's a sacred place to me. When I'm in the ocean, I feel as though I am part of the ocean.*
> —DAWN WRIGHT

But Dawn's participation in this dive is about more than checking an item off a personal bucket list. Her main motivation is the ambitious goal of mapping the entire ocean—70 percent of our planet, which remains largely uncharted. Although creative cartography may mislead us into thinking otherwise, only about a quarter of the seafloor has been mapped, to date, in high resolution.

On this mission, Dawn is acting as a flag-bearer for three initiatives:

- Seabed 2030, an international effort that aims to map every inch of the ocean in detail by the year 2030.

- Map the Gaps, a nonprofit that seeks to increase awareness, accessibility, and equity in ocean mapping.

- Adding data and maps from the deepest ocean to ArcGIS® Living Atlas of the World's extensive collection. ArcGIS Living Atlas is an ever-growing collection of authoritative geographic information, including maps, apps, and data layers, from around the globe.

During her expedition, Dawn (*left*) simultaneously promotes two ocean mapping initiatives with a banner and a T-shirt. She is pictured here with Rochelle Wigley (*right*), representing Map the Gaps, also a major flag-bearer for Seabed 2030 and the official mapper on Dawn's expedition, collecting and processing all the bathymetry for inclusion in the Seabed 2030 compilation.

Victor and Caladan are also fully engaged with these ocean-mapping efforts, having donated more than 1.5 million square kilometers (579,153 square miles) of bathymetric data collected over four years of expeditions.

And then, of course, there's the fact of who Dawn is: She's the first Black person and just the fifth woman ever to make the descent to Challenger Deep. She's acutely aware of the significance of that fact, and she hopes it will serve to inspire women and people of color to enter realms of science and exploration that were once inaccessible to them, by common practice if not by policy.

Dawn (*third from left*) poses with three other women who were part of Caladan's 2022 expedition: (*from left*) Nicole Yamase, the first Pacific Islander to make the descent to Challenger Deep; Kate Wawatai, a Maori New Zealander who is the first female pilot of the *Limiting Factor;* and Tamara Greenstone Alefaio, program coordinator for the Micronesia Conservation Trust.

Dawn is heartened, though, by the number of women who have made, and continue to make, enormous contributions to Caladan's efforts and to her dive in particular. But she notes that only seven of the 43 people aboard the *Pressure Drop* during this expedition are women. So, she says, there's still plenty of work to do, to ensure equitable access to science and technology education—and to encourage women and people of color, like her, to enter the field.

+0 HOURS, 20 MINUTES → +4 HOURS, 4 MINUTES
−1165 METERS → −10400 METERS

If not for its running lights, the *Limiting Factor* would be engulfed in total darkness for the remaining four-hour descent to the seafloor. The sub traverses three more vertical zones, each with an increasingly foreboding name: the midnight zone (down to 4,000 meters/13,123 feet), the abyssal zone (to 6,000 meters/19,685 feet), and finally, beneath that, the hadal zone, named after Hades, the Greek god of the underworld. This is the most tedious part of the adventure, but fortunately,

Dawn (*left*) and Victor at the controls of the *Limiting Factor*.

there's plenty to talk about. Some of it is critical to the mission, such as making sure Dawn is familiar with the basic controls of the *Limiting Factor* in the unlikely event that Victor should become incapacitated.

Dawn is well versed in submersibles, having done her PhD work using *Alvin* to study hydrothermal vents in mid-ocean ridges and *Pisces V* to study deep coral reefs in American Samoa. (Though they're both called subs, a submersible differs from a submarine in that it needs to be launched from a support vessel, whereas a submarine can launch itself and return on its own.) As a result, Dawn's able to settle in and approach a dive of this magnitude as she would any other, as a professional. With that mindset, she barely notices when the *Limiting Factor* floats past the 2,500-meter (8,202-foot) mark, making this now the deepest she has ever descended.

Compared with her shallower dives, the major difference in Dawn's preparation was a fasting regimen, since there's no latrine aboard the *Limiting Factor*. The round trip to and from Challenger Deep often takes at least 10 hours, so Dawn had to reduce her food consumption over two days, with a final snack and sip of water a few hours before launch.

Apart from that, there is little difference in the experience of the slow, dreamlike descent. If anything, it's more comfortable than her past dives, given the length of the dives the *Limiting Factor* has been designed for. Even the pressure outside the sub, increasing to almost unimaginable levels, is imperceptible to the occupants of the *Limiting Factor*. At the full depth of Challenger Deep, the pressure is 16,000 pounds of force per square inch, equivalent to a school bus sitting on top of every cubic inch of water (or to the atmospheric pressure on the planet Venus!)—but Dawn and Victor, inside the sub's meticulously engineered chamber, are protected from the ocean's crushing force.

Outside the hatch of the *Limiting Factor*, the crew have placed a handful of Styrofoam cups decorated with colorful doodles in a mesh bag. As the sub descends, the cups are fully exposed to the tremendous water pressure. When the sub returns to the surface, the cups are retrieved—warped and compressed to a fraction of their original size, but, amazingly, mostly intact.

One of the Styrofoam cups that rode all the way to Challenger Deep inside the hatch of the *Limiting Factor*, compared with a fresh Styrofoam cup.

Not even a year later, in June 2023, the tragic loss of OceanGate's *Titan* submersible near the wreck of the *Titanic* provided a sobering reminder of the perils of deep-sea exploration and the potentially catastrophic consequences of extreme underwater pressure. At the depth where the *Titan*'s hull suffered its deadly implosion, the pressure was an estimated 5,500 pounds of force per square inch, enough to crush a soda can to the size of a marble. Fortunately, unlike the *Titan*, the *Limiting Factor* has been constructed and certified to the highest industry standards.

+4 HOURS, 17 MINUTES | −10451 METERS

Apart from the expedition's symbolic importance, as a first for a Black woman, Dawn's dive has an immediate, tangible goal: pushing the science of seafloor data collection to new limits.

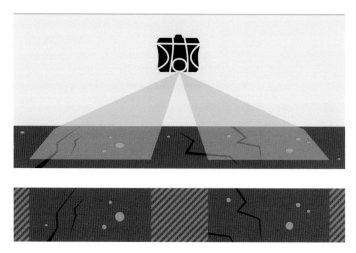

Above: An illustration depicting how a submersible like the *Limiting Factor* can obtain seafloor data using portable sidescan sonar. *Below*: How that data might be represented in two-dimensional form.

Multibeam sonar systems, such as the one attached to the bottom of the *Pressure Drop*, are effective for capturing the bathymetry (depth data) of large swaths of the ocean floor. Portable sidescan sonar systems, on the other hand, are used to produce more detailed images of the seafloor. These sidescan devices are often deployed closer to the seafloor and can more accurately read differences in material and texture down below. They do this by measuring the intensity of the return signals, rather than merely the time it takes for the signals to bounce back, as multibeam sonar does. This capability makes sidescan particularly effective for purposes such as finding shipwrecks, determining the state of underwater infrastructure, or locating mineral deposits.

But portable sidescan sonar has never been deployed deeper than 6,800 meters (22,309 feet); its circuitry generally doesn't hold up well against the immense pressure of the deepest sea. Now, though, a Mauritius-based company called Deep Ocean Search has developed a sidescan sonar apparatus that is designed to withstand the pressure and work at full ocean depth. Victor and Dawn have arranged for this new device to be attached to the outer shell of the *Limiting Factor*. If it functions successfully in Challenger Deep, it will represent a game changer in the field of seafloor mapping.

At about 10,450 meters (34,284 feet), still roughly half a kilometer above the floor of the trench, the moment of truth arrives. Cradling a laptop computer, Dawn powers up the sidescan for the first time. Much to her relief, the sonar's signals are reading clearly.

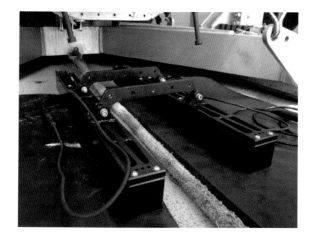

The sidescan sonar prototype developed by Deep Ocean Search that was attached to the bottom of the *Limiting Factor*.

Dawn smiles and gives a thumbs-up over success in getting live data from the prototype sidescan.

+4 HOURS, 28 MINUTES | −10904 METERS

There's hardly time to celebrate the success of the sidescan's operation before the bottom of Challenger Deep is upon Dawn and Victor. As the seafloor rises up to meet the *Limiting Factor*, they are greeted by a shiny, green, cylindrical object. "What is that?" asks Dawn, excitedly. "It looks like a light saber!" As they get closer to the object, they realize it's no natural formation.

"It's a freaking bottle," says Victor, with dismay in his voice. Indeed, it's a beer bottle, with the label intact. Victor's seen evidence of human life down here before—mainly gnarled cables, probably left behind by previous deployments of scientific vehicles—but it never fails to serve as a dispiriting reminder that human activity is affecting even the remotest spots on the globe.

A green glass bottle at the bottom of Challenger Deep.

As Dawn said later, "This is further evidence that we as humanity *must* do better by the ocean and the habitats that we ourselves share and ultimately depend on. There is no Planet B!"

But for Dawn, at least, the disappointment of stumbling upon the bottle is quickly replaced by a more positive emotion—elation. After all, she's made it to the deepest point in the world! It was, she said, a lifelong dream come true, a journey she'd never imagined would be possible.

"I don't know if I'll ever get to the moon," Dawn reflected after her dive, "but that was my moon walk, that was my moon shot. I think for all of us who have been to Challenger Deep, that is our one, that's the holy grail. And so I felt I understand now why astronauts are so poetic and they express themselves in a certain way that really gets to the heart of our spirituality and our existence, because they have seen Planet Earth in a way that we will never see it. And that's the way I felt in Challenger Deep. I felt astonishment and wonder and excitement."

Within a few minutes, the *Limiting Factor* reaches its maximum depth for this expedition: 10,919 meters (35,823 feet). Dawn and Victor take note of some holothurians (sea cucumbers) nestled on the seabed. It's something of an antidote to the bottle sighting; there's a certain comfort in realizing that forms of life have been surviving even in these most inhospitable places since long before our species evolved, and they'll likely continue to exist long after we're gone.

+4 HOURS, 43 MINUTES | −10890 METERS

Challenger Deep contains three distinct "pools" that drop below that magic 10,900-meter (35,761-feet) mark. The absolute deepest points are in the Eastern Pool, which has been the destination for most of the *Limiting Factor*'s descents. On this dive, though, Victor and Dawn are targeting the relatively unexplored Western Pool. In fact, their planned sidescan survey route will take them through completely new territory; they will be the first people ever to set eyes on the ground they're about to cover.

Their track takes them along the floor of the Western Pool and then gradually climbs the "wall" of the trench. All the while, the sidescan pings away. Its continued operation in this environment represents a massive technological victory.

But Dawn and Victor are making the most interesting observations with their own eyes. Now Dawn can see, firsthand, the results of geologic processes she's made a career out of studying. Before her lie fields of angular, blocky boulders, left in the wake of tectonic plates that have been colliding over millions of years.

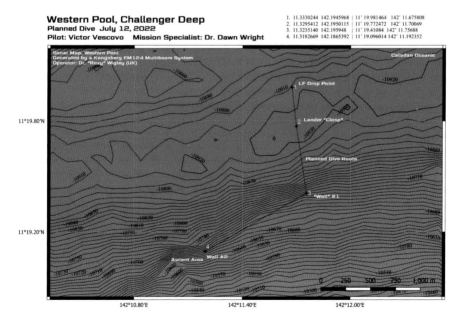

The bathymetric chart plotting the course Dawn and Victor took through Challenger Deep.

A field of boulders
left in the wake
of tectonic plates
colliding over millions
of years.

She christens one particularly impressive outcropping "Flintstones' Quarry." To explain its formation, Dawn later described how the seafloor is being "munched up" in this spot, right where "the Pacific plate is crunching into the Philippine plate and grinding up all these fantastically cool rocks."

+6 HOURS, 52 MINUTES → +9 HOURS, 53 MINUTES
−10727 METERS → −554 METERS

After nearly two and a half hours exploring Challenger Deep's Western Pool, the time has come to begin the ascent back to the surface. For all the *Limiting Factor*'s technological wizardry, this process simply involves discarding several metal weights. ("They'll decompose at this depth and become part of the seafloor," Dawn points out, though, sadly, that glass bottle never will.)

With the sub now appreciably less heavy, the return journey is about an hour shorter than the descent. Now is the time to decompress, both literally and figuratively. Victor's smartphone comes in handy here, as he's downloaded a couple of preselected movies to watch on the way up.

Dawn's pick is *Star Wars: Episode 1 – The Phantom Menace*. It might seem a curious choice, at least until she cites the memorable scene that takes place in an underwater vehicle. Then there's still enough time left for Victor to run through some handpicked scenes from his favorite film, *The Life Aquatic with Steve Zissou*. Before Dawn knows it, the water outside the viewports starts to take on a lighter hue. At last, after exactly 10 hours beneath the waves, the *Limiting Factor* bobs serenely to the surface.

6:11 P.M. | LAUNCH +10 HOURS, 0 MINUTES | DEPTH: 0 METERS

There's still plenty of daylight when the *Limiting Factor* peeks above the water, though the sun has migrated most of the way across the watercolor sky. A man in a fluorescent yellow vest dives into the ocean from a Zodiac raft and swims over to manually attach a tow cable to the submersible so that it can be brought back aboard the *Pressure Drop*. His name is Shane Muhl but he's affectionately known as Speedo.

Speedo's duty, which is not without risk, represents just one of many crucial roles that have enabled Caladan's missions over the past half-decade. These range from the engineers and technicians responsible for the groundbreaking technology that makes these expeditions possible, to the cartographers and scientists who can contextualize the expeditions' discoveries, to the crew that keeps the *Pressure Drop* running, including the onboard cooks.

Once pulled from the *Limiting Factor*, exhausted but still running on adrenaline, Dawn is more than ready for a hot meal prepared by those cooks. She has to keep her game face on a little longer to indulge the media crew that's there to document her achievement, but soon she's able to go to the restroom (!), shower, and join Victor in the galley for a plate of spaghetti Bolognese (his favorite postdive meal). It's past dinner hour for the rest of the ship, so the two explorers are able to share a quiet moment of reflection.

Shane Muhl stands on the submersible waiting for the *Pressure Drop* to hoist the *Limiting Factor* aboard.

Dawn *(left)* and Victor return triumphantly from the deepest point in the world.

For Victor, this may be the last time he'll ever pilot the *Limiting Factor* to the bottom of Challenger Deep, which makes the moment especially poignant. From the start, he's largely financed Caladan and its operations out of his own pocket, with the occasional infusion from "trench tourists" who are willing to pay a hefty fee to visit the deepest point on Earth. Now, though, he has found a buyer for the whole Hadal Exploration System—including the sub, the ship, and all the trimmings. Within a few months, video game magnate Gabe Newell and his organization Inkfish Expeditions will take over the operation, leaving Victor a bit nostalgic but eager for his next adventure.

For Dawn, the work continues—the essential work of mapping the deep and of explaining to audiences all over the world why it's so important to do so. Not just important, Dawn argues, but crucial to the future of our planet, about 70 percent of which lies under ocean waters.

As Dawn climbed out of the sub that day in 2022 and stood on the deck of the *Pressure Drop*, with TV cameras capturing her every move, scientists and nonscientists alike were celebrating her achievement. She even received a call from senior officials at the White House. People across the globe suddenly wanted to know more about Dawn Wright. Who was this Deepsea Dawn? What was her story? Why had she dedicated her distinguished career to ocean mapping? And how did she get to be the first (and so far, only) Black scientist to visit the deepest place on Earth?

Chapter 2

DEEPSEA DAWN

" *I was eight when I decided that I wanted to be an oceanographer, so it was, like, yep, this is what I'm going to do. Then it was just a matter of finding out, how DO you become an oceanographer?* —DAWN WRIGHT

The first time Dawn Wright saw the ocean, she was distracted. She was about three years old, and her parents, then living in Baltimore, Maryland, had taken her on a trip to Atlantic City, New Jersey. Although she admired the ocean, the future oceanographer was much more enchanted by the giant, animated Mr. Peanut on a billboard nearby.

But when Dawn was six and her family moved to Hawaii, the ocean became an essential part of her life. She spent her childhood immersed in the ocean and the ocean culture of the islands, and—inspired by the Jacques Cousteau documentaries she watched on TV—decided to become an oceanographer. The only question was: How? Unlike other professions, such as teaching or medicine, there was no existing academic track for her to follow.

Like most people's, Dawn's journey through life has involved unplanned twists and turns, moments when things didn't go as expected, or when a door opened and Dawn stepped through, not sure what would be on the other side. It wasn't always easy, being the only female or the only Black person in various academic and professional contexts along the way. Yet all these experiences, and Dawn's determination to make the most of them, have added up to a distinguished career, an adventurous life, and worldwide recognition as the scientist-explorer Deepsea Dawn.

Stepping out

If Dawn's path has involved a fair amount of stepping out, her mother, Jeanne (Grove) Wright, stepped out before her, forging her own path as a young woman. Dawn's mother was born in 1935 in segregated Baltimore, where the family had deep roots; Jeanne, the second of seven children, was the first to venture out of the area.

"She and her siblings were among the first to go to college," Dawn explains. "But she stepped out even further because she wanted to go to a liberal arts college in Illinois. So she stepped out in the 1950s to do that." First, Jeanne attended Howard University in Washington, DC. "But then she went all the way out to this little school, 25 miles west of Chicago, Wheaton College. And it was during the time in the Fifties when schools were not allowing African Americans to enroll. But this particular school, Wheaton, welcomed her and a small group of other African American students. And then she went from there to step out into South Dakota!"

At South Dakota State University, Jeanne completed a master's degree in language arts, but Dawn notes, "I don't think she ever saw another Black person the whole time she was there." As Dawn says, with pride, for a young Black woman to venture out like that was unusual in the 1950s, a testament to her mother's courage and drive.

Jeanne returned to the East Coast, began her career as an instructor of speech communication, and married Dawn's father, Robert Theodore Wright, an aspiring professional basketball player who became a high school coach. The family was living in Baltimore in 1961, when Dawn arrived on the scene: "I barely made it," she says, "because I was six pounds, one ounce, but I took care of that right away and just kept on growing."

When Dawn was five, Jeanne was offered a position at the University of Saskatchewan in Canada, so the family "trundled off" to Saskatchewan. Young Dawn barely had time to get used to the snowy northern plains and the friendly local kids before—"It was quite a shock"—the family moved again, this time to Hawaii.

Dawn in first grade, age six, Hawaii, 1967.

"So we went from Saskatchewan to Honolulu, Hawaii," Dawn says. "And we were in Hawaii for the next 10 years." The family lived for a year on Oahu, where Dawn's mother was teaching at the University of Hawaii at Manoa. Then she was asked to go to Maui to start the speech program at Maui Community College. Maui, in the 1960s, was "very rural," and Dawn believes her family was the first Black family to settle on the island, "so it was a new frontier." It was also the place where Dawn developed her passion for the ocean and for the volcanic rocks that surrounded her.

Hawaii

Hawaii was Dawn's happy place, the environment that formed her as a person and as an ocean geologist. Yes, she says, there were some "initial hiccups" when her family arrived on Maui—"I remember difficulty finding a place to live. There was some incident where someone didn't want to rent to us," as a Black family. And there was one bullying incident at school that she recalls as "very painful." But overall, Dawn says, "it just did not *matter*," because Hawaii was, and is, "such a crossroads of the Pacific." Dawn grew up with kids who were proud of their mixed ancestry—Japanese American, Chinese American, Filipino, Hawaiian, and Portuguese, to name a few—and the islands offered a diverse culture in which she thrived.

Growing up, Dawn spent much of her time at the beach with the other kids, playing, swimming, and "doing a little snorkeling and body surfing." She describes an adventurous childhood of "running on the beach, swinging from the vines of banyan trees, building things." And, she says, she became fascinated with rocks, all the different types of rocks and sand on the islands, while at the same time learning at school how those islands were formed. Even at a young age, the idea of all that undersea activity captured her imagination—volcanoes erupting from the seabed, seamounts breaking the surface to become the very islands on which she stood.

Something else that captured her imagination was the underwater photography of Jacques Cousteau. "I was a typical kid who sat in front of the TV every Sunday night at 6 o'clock," Dawn told an interviewer, "so I could get my weekly dose of *The Wonderful World of Disney*, followed immediately by *The Undersea World of Jacques Cousteau*, and then off to bed."

"That," she added, "was my steady diet of science communication!"

Even more riveting than Cousteau's voyages into the deep was the moon

Dawn in Hawaii in 1971, age 10.

landing of 1969. Dawn remembers watching the historic moment on TV, trans-fixed. After that, she says, "There were so many little kids running around with space helmets and wanting to become astronauts, and I thought about becoming an astronaut for maybe five minutes. But then I thought, 'Well, if those men could go all the way to the moon and do what they did, why can't I, as a little girl, go in the opposite direction, and aspire to become an ocean explorer?'"

Leaving Hawaii

By her freshman year of high school, Dawn was thriving in Hawaii, academically and socially, but then everything changed. Back in Maryland, her grandmother, Jeanne's mother, was in failing health, and Jeanne made the difficult decision to return to the East Coast to help care for her.

The move involved considerable sacrifice on both sides. Dawn's mother gave up tenure at Maui Community College, and the two of them had to "pick up stakes" and move across the country. By then, Dawn's parents had divorced, so Dawn and her mother stepped out "as a mother-daughter duo," at a time when single-parent families were not as common as they are today.

Pirates and the power of maps

Luckily, Dawn had a fourth-grade teacher who encouraged her to "read, read, READ!" and she did, devouring tales of adventure, especially those that involved the ocean. Dawn recalls being "completely enchanted" by Robert Louis Stevenson's *Treasure Island*, fancying herself as young Jim Hawkins, who befriended Long John Silver, the one-legged pirate. That early fascination with pirates has stuck with her, and Dawn is known these days for her love of 18th-century pirates and their lives of derring-do. She's even been featured as a pirate on a set of trading cards!

Four trading cards, front and back.

Beyond the allure of pirates, *Treasure Island* introduced Dawn to the allure of maps. "I realized the true power of mapping as an eight-year-old girl," she says, explaining that the novel's plot is driven by Captain Flint's treasure map—a map whose details are never revealed. "It is the sheer idea of the map's existence, and the possibilities of what it could lead to, that drives Long John Silver and the other characters to the point of obsession. For me, that was incredibly powerful!" Decades later, Dawn would still be championing the power of maps to tell a story, ignite the imagination, and inspire action.

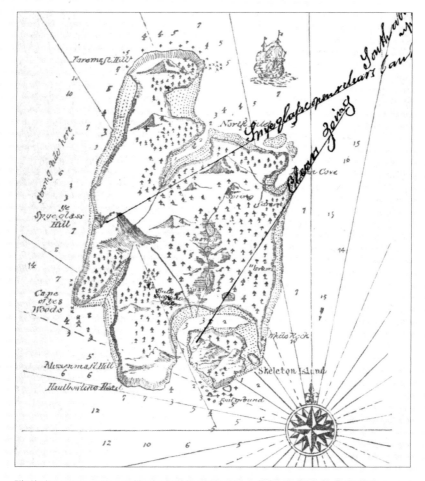

Flint's treasure map, as illustrated by Robert Louis Stevenson in the 1883 edition by Cassel.

Dawn says she was devastated by the upheaval: "It was a very hard move, because I did not want to leave Hawaii and we had a very good life there. She [Dawn's mother] didn't want to leave either, but this was a very important decision that she knew that she had to make."

6th ANNUAL MARYLAND MARATHON
DECEMBER 3, 1978

Dawn as a high school senior, finishing the Maryland Marathon, 1978, age 17.

Apart from the shock of leaving their lives in Hawaii, moving to the Baltimore metropolitan area meant culture shock for Dawn, a free-range Hawaiian kid. When she started high school in Columbia, Maryland, Dawn says, "I felt completely out of place, and I didn't feel connected to either White kids or Black kids." But Dawn was an athlete, active in basketball, track, and cross-country, and she was also highly motivated to study science. So sports and science helped her find like-minded friends.

An upside of the move was that Dawn's new high school was an innovative, "open space" institution, with a flexible curriculum that allowed students to pursue self-paced independent studies based on their interests and abilities. Plus, Dawn says, "I had some really fabulous teachers, including fabulous science teachers. And I also had the opportunity to do a science project with a professor at UMBC (University of Maryland Baltimore County), which really changed the course of my thinking about science. I was still interested in geology. And I did this science project on bioluminescent [marine] bacteria, which was fantastic." Dawn was "blown away" by the experience of doing actual research, petri dishes and all, which only heightened her desire to study ocean science.

But at the time, it wasn't possible to get an undergraduate degree in oceanography. (Today, aspiring oceanographers have a choice of undergraduate programs across the US.) So, Dawn says, "I knew I was going to head off to Wheaton College to be a geology major, to eventually find a program at the graduate level in oceanography."

Oceanographer-in-training

Why Wheaton? Dawn had always had her
heart set on Wheaton College, a small Christian liberal-arts college in Illinois, because
she'd grown up hearing about her mother's
fulfilling experience there. Although the geology program was small (Dawn was one of
seven majors), she received a good grounding
in the subject, while also being on the track,
cross-country, and basketball teams. This, for
Dawn, was the appeal of Wheaton: an environment in which she could flourish as a full
human being.

"I wanted to be at a school where I felt as
though I was able to develop as a whole person,
in mind, yes, but also body and spirit," she says:
"I wanted to be built as a good human being
first." This is a goal that doesn't always occur to

Dawn's senior picture, Wheaton
College, Wheaton, Illinois, 1983.

today's college-bound kids (and their parents), who are focused mainly on grades,
prestige, and future employment prospects. But it was crucial to Dawn, who says,
"I wanted to be able to mature in an environment that I felt comfortable in, that I
felt nurtured. There was no way that I was going to be able to get that, in my mind,
at a major, large university."

But a major, large university—Texas A&M—was exactly where she was
headed next, for her master's degree. For Dawn, this was another big step: going
from a campus with about 2,000 students to one that, at the time, had about
36,000. Although the campus of Texas A&M University is inland, at College Station, Texas, the school is known for its excellent oceanography program, focused
on the nearby Gulf of Mexico.

Dawn entered that program because she was offered a graduate fellowship,
so, like most grad students, she followed the money. Now, in retrospect, she says:
"Texas A&M was not a place that I thought I would end up in a million years, but it
turned out to be the right choice for me. It was a very, very good program, and the
Ocean Drilling Program (now known as the International Ocean Discovery Program) was going to come to the fore later, which was going to greatly affect my life."

And that would only have been available to me had I been at Texas A&M, not any other place. So it's really amazing how life is."

Though it worked out well in the end, Dawn's experience at Texas A&M definitely had its ups and downs. When she went on her first student cruise on a research vessel, she was horribly seasick, "flat on my back in my bunk," for three out of the five days at sea. "I reconsidered whether I was in the right major," she admits, but she soldiered on, seasickness and all.

Dawn confronted a more serious setback when she submitted her master's thesis, "The Nature of the Northern Terminus of the Tonga Trench, Southwest Pacific." This was a geophysical thesis, a gravimetric study of the Tonga Trench using seafloor gravity data and bathymetric data that had already been collected. Her major professor, who led her thesis committee, hadn't had much time to advise or support her, but Dawn did her best and the committee grudgingly passed her. When the professor told Dawn that she'd passed, he also told her that, basically, she didn't have what it took to be an oceanographer. Perhaps she should consider another profession? Go to law school? Get an MBA?

Failure is not final

Of course, her adviser's reaction was a "huge blow" to Dawn at the time. She understood that, as an assistant professor fresh out of graduate school himself, he was under enormous pressure to "publish or perish," and so Dawn, as his advisee (and his very first one at that), had been more or less left to fend for herself. "I had to sink or swim on my own," she says, "and I found resources where I could, mainly with the graduate students in the department of geophysics," who taught her what she needed to know to complete her thesis. With their help, Dawn was able to write it and defend it, without much help or input from her major professor.

When he delivered his devastating assessment, Dawn admits, "I felt very deflated and distressed. *But*, at the same time, I had passed. I had gotten that master's degree. And I had also heard at that time, the Ocean Drilling Program was looking for marine techs. And I had applied and been accepted. So I knew that I had that as a next step. And that I was going to learn more and more about oceanography. I was going to get this amazing experience. I was going to be able to go to sea into the field and learn more than I could ever have imagined, even more so than going directly into a PhD program. So even though I was discouraged, I thought, well, that just doesn't make sense. There's no way in this world that I'm going to go to somebody's law school! I don't want to be a lawyer. And I am not a business person. That just doesn't compute, and I'm not going to accept it!"

And so, as Dawn always says, "This is the story of *failure is not final*. Failure is not fatal, and therefore, it's not final." It certainly wasn't for Dawn, who persisted on her chosen path, and, proving her adviser wrong, became not just an oceanographer but a world-renowned one—an elected member of the National Academy of Sciences and the National Academy of Arts and Sciences, among numerous other honors, and the author/coauthor of 13 books (to date) and hundreds of articles. Not to mention one of the select few scientists invited to visit Challenger Deep.

First, though, she had to spend a few years at sea.

Gap year(s) at sea

Around the time Dawn was finishing up her graduate program at Texas A&M, the university was awarded a huge contract to oversee the science operations of what Dawn calls "one of the most amazing earth science programs ever." This program, which originated in the 1960s as the Deep Sea Drilling Project, was a multinational collaboration that aimed to study and reconstruct the Earth's history by drilling samples from the rocks and sediments of the world's oceans.

When that program came to an end, scientists pushed for the drilling project to continue, so it was reestablished and funded as the Ocean Drilling Program, with more nations involved and a brand-new drilling vessel—the *JOIDES Resolution*. JOIDES stands for Joint Oceanographic Institutions for Deep Earth Sampling, while *Resolution* is a tribute to the ship in which Captain James Cook explored the Pacific.

"So all of this was incredibly exciting," Dawn says, "and, for me, it was perfect timing because I had just gotten my master's degree." She had been offered a position as a marine lab tech aboard the new drilling vessel, which, she says, was a wonderful opportunity. The ship, outfitted with various types of laboratories, would travel from point to point across the oceans, carrying an international group of 20–30 scientists and an additional 50–70 crew members, including the scientific technical staff.

For the next three years (1986–1989), Dawn would sail around the world (seasickness and all) and gain invaluable hands-on experience in geologic oceanography. She describes it as "a lot of time at sea, a lot of experiences with the different kinds of scientific projects that were being undertaken" and the different

Dawn (*center, near back*) as an
Ocean Drilling Program marine
technician, aboard the *JOIDES
Resolution* drillship in the Weddell
Sea, Antarctica (leg 113), 1987.

questions the scientists were exploring. Dawn worked 12-hour days, helping to
bring up the sediment and rock samples, processing them in the labs, and managing the data. For her, it was an opportunity that no graduate program could have
provided, "getting exposure to all of these top-level scientists from all over the
world, also getting exposure to the men who were working, the crew, the drilling
crew, the ship's captain, the mates, the engineers, and so forth." Overall, Dawn
says, "it was just a fantastic experience."

It was also good training for the (at the time) male-dominated world of ocean
science. Aboard that vessel, she told an interviewer, "There were 100 souls, and
10 of us were women. I think there were always a few women in the scientific
party. There were several of us women who were technicians. So we were working
in the labs, running the core samples through the various tests, cataloging and
photographing, doing all of that." But, she notes, "this was a drilling vessel that
had been refitted from a standard oil-and-gas exploration vessel, totally a man's
world." So, in Dawn's words, "there were varying levels of acceptance" of the

women onboard. Despite these divisions, Dawn made some good friends among the crew: "I've found that when you give yourself an opportunity to just sit down with people, you can relate."

Lassoing icebergs

To drill the samples from the ocean floor, the *JOIDES Resolution* had to be anchored at a single spot for days at a time. In the Antarctic, in the stretch known as Iceberg Alley, this posed a significant risk—of being hit by an iceberg! So the research ship had to have a support ship, which, as Dawn says, "was basically there to lasso icebergs and drag them out of our way."

This deepwater towing and supply ship (the *Maersk Master*) was Danish, and Dawn says, "I'd never met a crew of sailors who were so excited about their job. They just thought that they were the best because they were basically there to keep us from danger. They were like superheroes on the high seas."

The support ship followed Dawn's expedition at sea for the entire two months and, when an iceberg was spotted, would calculate the risk it posed to the ship (considering the winds, the iceberg's speed, direction, and other factors).

Then, Dawn says, "when an iceberg would get into the danger zone, into this buffer zone where it might affect us, they would steam up to that iceberg. They had this huge, humongous rope. It was basically a rope that they would drop over the side, right at one point of the iceberg. They would steam around, letting out the rope as they went, until they encircled the iceberg, and they would pick up that rope and then, literally, like a tugboat, tow the iceberg out of the way." This was an operation that could take several hours, depending on the size of the iceberg (one was the size of a football stadium).

One intrepid Danish crewman took the leavings of the ancient sediments that the scientists had drilled from the depths and fashioned a clay penguin out of them. This penguin, fired in a little furnace and painted just for Dawn, remains one of Dawn's most cherished possessions. For her, it's a beautiful token of cross-cultural friendship and a lasting reminder of the seaman's talent and creativity.

Dawn with the clay penguin made for her by a crew member of the Danish deepwater towing and supply ship, the *Maersk Master*, which accompanied the *JOIDES Resolution* in the Antarctic, 1987.

As a technician, Dawn worked a schedule of two months at sea, then two months off, for 10 consecutive expeditions in the Indian and Pacific Oceans. Along the way, she got to see some far-flung parts of the world—Senegal, Barbados, Antarctica, Chile, Mauritius, Kenya, Australia, New Zealand, Singapore, the Philippines, Tokyo, and other desirable destinations. As Dawn says, "For a young person just coming out of graduate school still wanting to work in her field, this was just fantastic, getting to travel the world along with it."

Through her resilience, her determination, and her work ethic, Dawn had created an exceptional opportunity for herself—a three-year gap year, of sorts, before gap years were a thing. The experience she gained, in science and in life, was priceless. But Dawn was still focused on her goal of becoming an oceanographer. It was time to jump-start the next phase of her life. Time to go back to school.

Enter GIS

Although she had a master's degree in oceanography and had been working on a geologic expedition for three years, Dawn's next step was to apply to geography programs for her PhD. Why geography? Well, Dawn says, this is where GIS—the spatial technology in which she would eventually specialize—enters the picture.

One day, in the late 1980s, Dawn's mother sent her a tiny but intriguing clipping from *The Chronicle of Higher Education*. It reported that the National Science Foundation had funded three universities to establish nodes of the National Center for Geographic Information and Analysis (NCGIA) on their campuses. Dawn's interest was immediately piqued. At the time, GIS "was a very exciting field that was opening up," and Dawn began to consider geography as an integrative discipline that she could embrace: "I could still study the oceans as a geographer, but this whole new world that was opening up with mapping technologies and spatial analysis, that really attracted me."

The University of California, Santa Barbara (UCSB), was one of the institutions with an NCGIA, and, when Dawn applied there, she was offered a President's Fellowship to support her doctoral studies. That sealed the deal, and in 1990 Dawn was off to California.

Once again, she had landed in the right place. Although Dawn had entered a geography program, UCSB also offered an Individual Interdisciplinary PhD Program, where a student could pursue a research question that could be addressed only by integrating two or more disciplines. So her doctoral studies in geography expanded to include geologic oceanography, with the support of researchers from UCSB's Marine Science Institute. And this time (unlike with her master's degree) she lucked out with her thesis advisers.

As Dawn explained in a later interview for UCSB, "[Dr.] Ray Smith (Geography) was not an expert at all in the subject of my dissertation but was absolutely the very best major adviser that I could hope for due to his care and skillful mentoring (dispelling the myth that graduate students need to be clones of their major advisers). [Dr.] Rachel Haymon and [Dr.] Ken Macdonald (Earth Science) gave me absolutely invaluable experience at sea, wonderful mentoring, and welcomed me fully into their research labs. Many of the students in those labs are close friends of mine to this day."

Dawn's dissertation research was focused on making maps of the ocean floor with this relatively new technology known as GIS. As she describes it, her goal was "to understand more of what we now know as data science issues—how we collect and manage and interpret data from the seafloor," increasingly in real time. The data she wanted to interpret was not only geological but also biological and chemical. So, she says, "enter GIS, which at the time was very good at managing different types of data, but which had not yet been applied very much to the ocean."

Through her collaboration with marine geologist Rachel Haymon, Dawn had the opportunity to work on one of the first GIS datasets ever collected by the deep-sea camera sled *Argo II*. (*Argo*, named for the mythical vessel that bore Jason in his quest for the golden fleece, was also famously used by Dr. Robert Ballard to discover the wreck of the *Titanic* in 1985.) Haymon was using the *Argo II* to study hydrothermal vents—underwater hot springs—on the ocean floor south of Acapulco, Mexico. The technical crew was experimenting with different technologies, including GIS, and Haymon received her data in GIS format, which, at the time, she knew nothing about.

Enter Dawn, who very quickly got up to speed with GIS and began to work with the data. "It was absolutely thrilling," she says. "I had never been so happy to be locked in a small, windowless lab for hours on end as I was witnessing on video and trying to wrangle data-wise what few others at the time had ever seen."

Haymon also involved Dawn in her research expeditions to a segment of the East Pacific Rise where two tectonic plates are pulling apart; this area, notable for its volcanic activity and hydrothermal vents, became Dawn's area of study. And, to study this geologic activity close up, Haymon had received funding to use the DSV *Alvin*, a crewed deep-ocean research sub owned by the United States Navy and operated by the Woods Hole Oceanographic Institution.

Dawn's experiences with *Alvin* in 1991 were landmarks in her career—and, indeed, in marine geology. For starters, on the first expedition Dawn joined, Haymon and her crew came the closest that any scientist had ever come, at the time, to witnessing a volcanic eruption on the ocean floor. Unlike volcanic eruptions on land, which are hard to miss, eruptions on the seabed happen in the obscurity of the vast ocean. "It was just serendipity that brought us to within two weeks of actually seeing a volcanic eruption," Dawn says. By chance, Haymon was there to see "the aftermath of an eruption that had occurred two weeks earlier, and we know that from the dating of the volcanic ash. So it was a very historic expedition."

The expedition was historic for another reason, too. On the first dive of that cruise, Haymon went down in *Alvin* with another female scientist, Dr. Karen Von Damm, and a female pilot, Dr. Cindy Lee Van Dover. "That was the first time that there had ever been a female pilot of *Alvin*," Dawn notes, "and with two women scientists, the first time there was an all-women crew in *Alvin*."

Then Dawn herself made history. She became the first Black woman to dive to the ocean floor in *Alvin*, reaching a depth of 2,500 meters (8,202 feet). As well as being historic, that dive was an unforgettable, magical milestone for Dawn the scientist and Dawn the human being, entering an alien world for the first time.

First dive on *Alvin*

So my first dive on *Alvin*, it just blew my mind because I had always wanted to do any type of submersible dive. Right before I took my dive, one of my colleagues said to me, "You know, you're going to be the first African American woman to dive in *Alvin*, this is going to be a pretty big deal," and I thought it shouldn't be such a big deal. We want to get a much broader cross section of people in the sciences to have these really special experiences.

When you are in *Alvin*, and you slowly descend, it's just a beautiful, calm, amazing feeling. There is this period where you descend from where you can see in the ocean, which we call the euphotic zone, so it's a beautiful blue, but then it starts to get gray and then it goes pitch black, so it's like you are in space. We started to see these flashes of light in the water, in the darkness, and they were bioluminescent bluish-green siphonophores, or worms, that were bumping into each other and twirling around, and it was like a fireworks show right outside of our windows. Finally, after a couple of hours we got to a depth of around 2,500 meters, so that's about a mile and a half, when the pilot first turned on the lights. He said, "We've landed, we've touched down on to the seafloor"— just mind blowing, because there it was, there was this very strange alien volcanic seascape. Everything was completely still.

Dawn diving in *Alvin*.

Certainly, you can cover a lot more of the seafloor and the water column with these other types of robots that are autonomous but there's still something about being there, about the perspective and the awareness, and even the sense of scale. The tubeworms that we saw were three to four feet long, the clams that are at these hydrothermal systems are the sizes of dinner plates.

Pretty much everything that we know about these underwater hot springs is because of submersibles like *Alvin*—you move along the seafloor from target to target. For me, it was looking at the shapes of the lava flows, because my emphasis was looking at the distribution of cracks on the seafloor, or fissures, and then, of course, making all the observations that we could in terms of the hydrothermal vents.

Until it's time to slowly rise up to the surface.

We were so tired that we basically slept on the way back up! And then you're brought back on board, and it's just a really amazing, exhilarating experience, and there's nothing like the first *Alvin* dive.

Professor Wright

Right around the time she was completing her dissertation, and just when she'd been invited for a job interview at Oregon State University (OSU), Dawn had to be helicoptered out of a ravine. She'd been in a mountain biking accident—flew off a mountain, landed 30 meters (100 feet) down a ravine. This was not exactly the type of dive that Deepsea Dawn was looking for.

Luckily, she hadn't broken any bones, although she sustained multiple injuries and had to have knee surgery. "I'd come very close to being killed, actually," Dawn says, but at the time her focus was on "trying to get healed up from that and to finish writing my dissertation and to just move on from there."

Dawn's dissertation, titled "From Pattern to Process on the Deep Ocean Floor, a Geographic Information System Approach," had attracted attention from OSU because the (then) Department of Geosciences was looking to hire someone who had expertise in both geography and geology—which Dawn most definitely did. Her dissertation was a study of hydrothermal vents on the ocean floor in the East Pacific Rise. Thanks to *Alvin* and *Argo*, Dawn says, "We were able to look at geological processes but also biology, where different creatures were living at these vents, the chemistry of the vents, and how the vents were distributed on

Dawn (*right*), PhD, who graduated from UCSB in December 1994, attends her June 1995 commencement. Her mother, Jeanne Wright, presents a lei.

the seafloor." Using spatial analysis and GIS technology, she was able to integrate these different kinds of data and interpret them. It was, she explains, "a combined story of geology, biology, and chemistry, to understand those hot springs on the ocean floor."

Once Dawn was able to hobble up to Oregon for her interview, she was offered the job: Assistant Professor in the Department of Geosciences at Oregon State University, Corvallis, Oregon. (Not bad for someone who "didn't have what it took" to become an oceanographer ...) But Dawn was reluctant to move into a tenure-track faculty position without first doing postdoctoral research, as is the norm. So she negotiated with OSU to do an eight-month postdoc with her colleagues at the NOAA Hatfield Marine Science Center, an extension of the OSU campus, continuing her work on hydrothermal vents.

When the postdoc was complete, in 1995, it was time to begin her teaching career in earnest. Dawn had a typical—i.e., heavy—workload of teaching, research, and service, the "service" component being larger than usual because she was skilled in GIS, which was rapidly growing in popularity. As an early adopter, Dawn was called upon to help coordinate GIS across the campus and revamp the GIS program. Though she flourished personally and professionally

Dawn as a new assistant professor at Oregon State University, with her puppy Lydia, 1996.

at OSU, Dawn points out that, for most of her time there, she was the only Black female faculty member in the entire College of Science.

As she made her way through college, grad school, and professional life, Dawn had to find her own mentors—and she found wonderful ones—from different backgrounds, because there was no one to turn to who looked like her. But, she says, "If I had limited myself to only wanting to see role models or mentors who looked like me, I wouldn't have had much to look toward. So I think another message is to find your inspiration, be willing to find your inspiration in all different ways through different kinds of people. Now I love being a role model and a mentor to young people, including African American people who want to consider going into the earth sciences or studying the oceans. I love that. I love that there are more of us now."

> *We desperately need the perspectives of more people of different cultures and ethnic backgrounds, men and women. We need this to guide our educational programs, our research, our governments, and our GIS companies!*
> —DAWN WRIGHT

As a testament to Dawn's years of teaching and mentorship, she was recognized as a United States Professor of the Year for the State of Oregon by the Carnegie Foundation in 2007. That, she says, was a "real shocker," because she didn't think she had a natural gift for teaching. "Teaching is very hard for me," she says, "and I've had to work incredibly hard, and I used to be scared to death in the classroom. But over the years, it's been, I guess, a labor of love to come into my own as an instructor."

Along with the teaching, Dawn continued to produce reams of research. In

1996, she returned to the Tonga Trench (site of her ill-fated master's thesis) for a mapping project, this time with a happier outcome. Collaborating with her department chair at the time, Sherman H. Bloomer, Dawn mapped the trench between Fiji and Samoa, and was also able to map the second-deepest spot on Earth, Horizon Deep (10,823 meters/35,509 feet). She could not possibly have imagined that, over two decades later, she'd be descending to the only point on the planet that's even deeper.

In 2001, inspired by the distinguished oceanographer Sylvia Earle and her Sustainable Seas Expeditions—an initiative to map all the US marine sanctuaries—Dawn took a sabbatical year to map the (then) Fagatele Bay National Marine Sanctuary in American Samoa. American Samoa, a small archipelago of five islands and two coral atolls, is part of the United States—which many people don't realize—and the only US territory south of the equator. It's a region where Dawn's research would produce tangible results.

Four years later, Dawn returned to Samoa, this time as chief scientist for an expedition that used the *Pisces V* submersible, similar to the *Alvin*, to identify new species of coral and invertebrates. Dawn made two dives in *Pisces V*, owned by the Hawaii Undersea Research Lab, and also worked closely with the local American Samoan community in its efforts to monitor and protect the coral. Thanks in part to this outreach, and to the GIS toolkit that Dawn's team built, the marine sanctuary was expanded and renamed the National Marine Sanctuary of American Samoa.

Over the years, Dawn's fieldwork has taken her to some of the most remote and geologically active regions of the planet. You would need a high-resolution seabed map to plot all the locations where she's conducted research—Samoa, the Tonga Trench, the East Pacific Rise, the Mid-Atlantic Ridge, the Juan de Fuca Ridge, volcanoes under the Japan Sea and the Indian Ocean, the Mariana Trench … the list continues to grow. But, even with all this globetrotting and growing international renown, Dawn had a secure home base at OSU, where she'd been tenured since 1999.

With a permanent position at a research institution where she was productive and highly valued, Dawn seemed to be settled. And yet, in 2011, she made the risky decision to "pick up stakes" again, this time for a position outside academia and a base in Redlands, California: headquarters of Esri. Once again, a door had opened, and Dawn stepped through it.

Chief Scientist of Esri

As a researcher and practitioner of GIS, Dawn had long been using Esri technology, along with other technologies, in her work. As a professor, she'd taught her students to use Esri tech, because Esri is the global market leader—or what Dawn calls "the 800-pound gorilla"—in GIS software, location intelligence, and mapping.

But as an advanced user and oceanographer, Dawn was also a squeaky wheel. "I was sort of knocking on Esri's door," she says, advising the company that its products were "not what they should be" for the oceanographic community. Dawn saw a gap between how GIS was developing and how it could be used more effectively for science, including ocean science. Recognizing the urgent need to map and protect the ocean, which requires a rigorous, data-driven approach, the company took Dawn's concerns seriously.

Over the years, Dawn (the squeaky wheel) began to work more closely with Esri, regularly attending its annual user conference and getting to know the maritime team. At the same time, Esri was becoming more focused on developing technology for ocean science and ocean resource management. In 2001, Dawn, as an Esri customer, joined forces with Esri employee Joe Breman to create an Ocean Special Interest Group for the Esri User Conference. "So that's actually how that started," Dawn says—and the rest, as they say, is history.

After many years of working closely with the company, Dawn was invited, in 2011, to consider a position as Esri's chief scientist—and to lead Esri's new initiative in applying its technology to the oceans. The timing was fortuitous, because Dawn was about to start a sabbatical and "was looking for a bit of a change of direction anyway." As part of her sabbatical, she arranged with OSU to work at Esri full time for two years. Her intention, in taking the position, was to return to OSU when the two years was up. But somehow that never happened. ...

Leaving academia for industry was not a decision Dawn took lightly, but she was persuaded to stay on at Esri, giving up her tenure at OSU and switching her status to courtesy or affiliated professor. "I have not regretted taking this step away from the ivory tower," Dawn wrote in 2013. Nor, she emphasized, had she ceased to be an academic and a researcher, as required by her new role—still publishing, still attending conferences, still giving lectures and workshops, still reviewing papers and proposals, still active in scientific policy, but without the burden of having to "constantly chase after grant money."

At Esri, Dawn found herself in the enviable position of being able to work

Dawn on the Esri User Conference plenary stage, 2017.

on dream projects for ocean science—3D and 4D data visualization, for instance, and complex fluid systems modeling—while also having a wider reach as a liaison and ambassador for science. Ocean science is part of her mandate as chief scientist, but, she says, "I always remind people that I was never hired as Esri's chief *ocean* scientist."

Dawn represents Esri to the international science community at large—in disciplines ranging from climatology to conservation biology to geodesign—while working to strengthen the scientific foundation for Esri software and services. At the same time, she's helped to develop a flourishing ocean team at Esri, with an expanding portfolio and stronger linkages to NOAA and NASA. These days, it's rare to catch Dawn on a day when she doesn't have a conference to attend, a podcast to record, a plane to catch, a presentation to give, research to write up, young people to mentor, or an inbox of emails requesting more of the same.

Yet, even with this high-profile position and all the time, energy, and travel it requires, Dawn has managed to remain true to who she is: "Deepsea Dawn," a renowned scientist and explorer; a lover of LEGO, Snoopy, *Star Wars*, and Sponge-Bob SquarePants; a teacher and mentor; a voice for equity and inclusion; and an advocate for a more humane approach to science.

Dawn matter-of-factly points out that, for most of her undergraduate and graduate career, she was the only one—the only Black person in the room, often the only woman. But, she says, that is changing—if not quickly enough—with many more women in science these days, and more role models, such as herself, for young people from diverse backgrounds. She has often said that she looks forward to "a world where we don't have to keep revisiting the same problems that we thought were long since solved," including various forms of bias, "and as a result can truly move forward."

> " *I think my legacy is one of stepping into these new frontiers, these new territories, and I hope that others follow suit because it's really fun and exciting, and we can do it. We can do it.* —DAWN WRIGHT

Dawn has faced, and overcome, many challenges in her life. She has flown high in her career and dived deep. As Deepsea Dawn, she descended three times in *Alvin* and twice in *Pisces V* to study the shape of the seafloor. But, for most of her career, there was one challenge, one dive, that seemed out of reach—Challenger Deep.

Before 2019, only three humans had ever visited it. No known vessel could reliably withstand the crushing pressure at Challenger Deep. It remained a tantalizing mystery, the deepest, darkest, coldest, loneliest spot on the planet, more difficult to reach than the surface of the moon. How could Dawn ever have imagined that, in 2022, she would dive there herself?

Chapter 3

THE MYSTERIES OF CHALLENGER DEEP

" *I want to go to the deepest part of the ocean. I mean, who doesn't?*
—Dr. Sylvia Earle

C hallenger Deep is … deep. Located within the already-deep Mariana Trench in the western Pacific Ocean, it's the deepest point in the world ocean, the deepest point on planet Earth. Unless you are one of the lucky few, like Dawn Wright, who has actually descended there, the extreme depth of Challenger Deep strains most people's imagination. Perhaps we can picture the underwater wonderland of a scuba dive or nature documentary—but the dark, fathomless depths of the hadal zone? Probably not.

So let's explore some ways to visualize this depth, those seven long miles of Dawn's journey into the unknown. First, though, we need to understand why Challenger Deep is, in fact, so deep.

The Pacific Ocean is massive; the whole of it cannot even be seen from space. Deep below the coasts of the Pacific, the tectonic plates of Earth's crust grow, spread, compress, drift, and otherwise grind against their neighbors. The Pacific Plate is the largest of these fractious rock shells, floating atop a roiling sphere of magma.

Along its eastern edge, the Pacific Plate is slowly spreading away from, or sliding along, its neighbors. Much of the western edge, containing some of the oldest material of Earth's crust and therefore comparatively dense and cool, is colliding with its neighbors under incessant force, sliding below those plates in a process

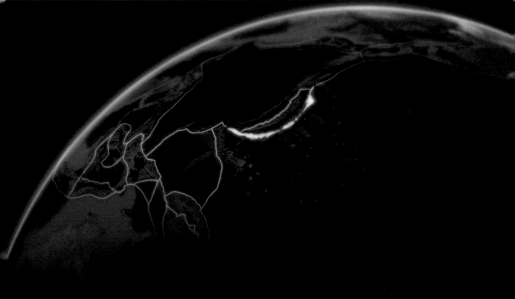

The Pacific Plate.

known as subduction. Over time, subduction zones like these can trigger phenomena such as earthquakes, tsunamis, and volcanic eruptions. In this case, the underthrusting of the Pacific Plate beneath the younger and lighter Mariana Plate has formed a deep crescent-shaped wrinkle where they meet.

This is the Mariana Trench: a 2,550-kilometer-long (1,580-mile) groove that plunges to the greatest depths on Earth. It runs parallel to an archipelago, the Mariana Islands, also formed during the processes of subduction. And at the southern end of the Mariana Trench lies a relatively small slot of even greater depth: Challenger Deep.

A closer look reveals that Challenger Deep is made up of three relatively small basins, or "pools"—Western,

Tectonic plates in yellow with arrows pointing to where the Mariana and Pacific Plates meet. The blue glow highlights the Mariana Trench.

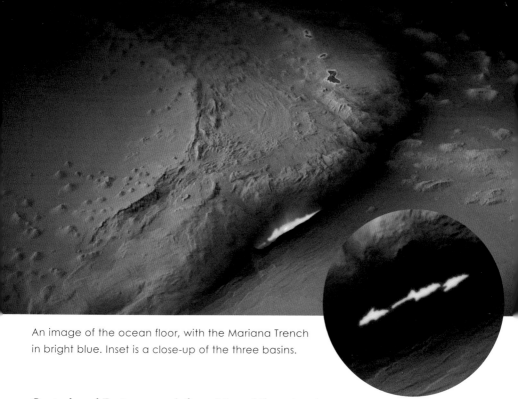

An image of the ocean floor, with the Mariana Trench in bright blue. Inset is a close-up of the three basins.

Central, and Eastern—each from 6 to 10 kilometers (3.7 to 6.2 miles) long. The relatively unexplored Western Pool was Dawn and Victor's destination on their historic dive.

When you consider what a tiny wrinkle Challenger Deep is in the vastness of the oceans, it makes you wonder: How did we ever find it? What were the odds of humans happening upon the deepest spot on Earth? And how on earth did we manage to measure it?

The answer is, purely by chance. In 1873, HMS *Challenger*, a small British Navy warship converted into the world's first dedicated oceanographic vessel, embarked on an eastward circumnavigation of Earth. Aboard were John Young Buchanan and John Murray, naturalists and pioneers in the relatively new field of oceanography.

Two years into the scientific expedition, the *Challenger* attempted to make landfall at the island of Guam but was blown west of its destination. The team took the opportunity to make "soundings" (lowering a weighted rope to determine the depth of the seafloor) along this unplanned track and chanced upon the erratic depth of the Mariana Trench, only 15 miles from what is now known as Challenger Deep.

Almost a century later, in 1960, the *Trieste*, a small free-diving submersible craft crewed by US Navy Lieutenant Don Walsh and Jacques Piccard (the son of

A painting of HMS *Challenger*, 1858.

The bathyscaphe *Trieste*.

the engineer who designed the *Trieste*), made the first crewed descent to Challenger Deep. Just below 9,000 meters (29,527 feet), one of the outer acrylic viewing panes cracked, shaking the entire vessel (and no doubt the crew). Undeterred, Walsh and Piccard continued with the five-hour descent but spent only 20 minutes at the bottom, keeping an eye on that potentially catastrophic crack.

Fast-forward to 2019, when regular dives in Victor Vescovo's DSV *Limiting Factor* began—and the rest is history, some of which was made by Dawn Wright on July 12, 2022.

But how can those of us sitting comfortably in our chairs at home even begin to imagine a dive into those depths? You could say that Challenger Deep is equivalent, in miles, to a walk down Manhattan from 110th Street to the World Trade Center … but the path to Challenger Deep is not a leisurely stroll through Central Park. It's a plunge through absolute darkness and crushing depths.

So let's try a thought experiment. Imagine you're sitting in a boat directly over Challenger Deep, and a hole opens up right there. An impossible, imaginary shaft all the way down to the bottom.

Now imagine you are a skydiver, but instead of jumping from a plane, you jump from the boat and begin your free-fall plummet to the bottom.

Your speed would rapidly increase until you were about 450 meters (1,476 feet) down, at which point you would hit terminal velocity and hold steady at a screaming 195 kilometers per hour (122 miles per hour).

You would continue to rocket toward the bottom of Challenger Deep for the longest three and a half minutes of your life. This is enough time to listen to the entirety of the Beatles' "Across the Universe" and still have four seconds remaining to pull your chute and land safely.

But, of course, there is no convenient shaft of open air from Challenger Deep to the surface. So, instead, how about swimming?

Not just any sort of swimming, though. Consider the extremely dangerous world of competitive no-limits free diving. Divers hold their breath and plunge to enormous depths and resurface, vying for the title of deepest. The current world record for a no-limits free dive is held by Herbert Nitsch at a lung-macerating 253 meters (702 feet) …

… which, in the context of Challenger Deep, looks like a tiny blip.

When he ascended from this record depth, Nitsch experienced several strokes due to decompression and required an immediate medical evacuation and extensive rehabilitation.

To reach the bottom of Challenger Deep, a diver would have to keep going, repeating Nitsch's perilous record-setting depth—already at the edge of humanity's capacity to survive—another 49 times. Inconceivable.

Then consider the pressure down there. The weight of the ocean at Challenger Deep squeezes from all directions, exerting a pressure more than 1,071 times the pressure at sea level. A one-liter bottle of air carried to Challenger Deep would be crushed to the volume of a pea. A deep breath would be compressed to the volume of five peas. Nearly nothing.

For the opposite scenario, picture a climber on Mount Everest, struggling to breathe or relying on supplemental oxygen. One of the greatest challenges of summiting the world's highest peak is that the atmosphere and oxygen levels at that elevation are dangerously low. But even the highest place on Earth would be swallowed up by Challenger Deep. If we could somehow plop Everest into Challenger Deep, it would still lie under more than 2 kilometers (1.2 miles) of water and be shrouded in absolute darkness.

In fact, sunlight capable of supporting photosynthesis only penetrates to the topmost 200 meters (656 feet) of the ocean. This zone, known as the euphotic zone or sunlight zone, is home to microscopic phytoplankton responsible for producing half of Earth's oxygen. It's strange to think that every other breath we take comes from the light-bathed microscopic organisms here. But it also reminds us how crucial the health of the oceans is to all life on Earth.

Beyond 1,000 meters (3,280 feet), absolutely no sunlight penetrates—not a single particle. Save for the rare bioluminescent pulse of sea life, most of the descent to Challenger Deep takes place in utter darkness. The absolute maximum distance that light penetrates would have to be traversed another 10 times to reach Challenger Deep.

Unless we're adventurers or adrenaline junkies, it's probably hard for most of us to relate to free diving, skydiving, bone-crushing pressure, and absolute darkness. You may not have visited Mount Everest either, but perhaps you've seen the Grand Canyon?

Its scale can be dizzying when beheld from the rim. The Grand Canyon, at 1,857 meters (6,093 feet; over 1 mile), is so deep that it has a different climate at the bottom.

Now imagine that we could magically scoop out the Grand Canyon, with a giant spoon, and use it as a handy unit of reference …

… it would only dip into the midnight zone.

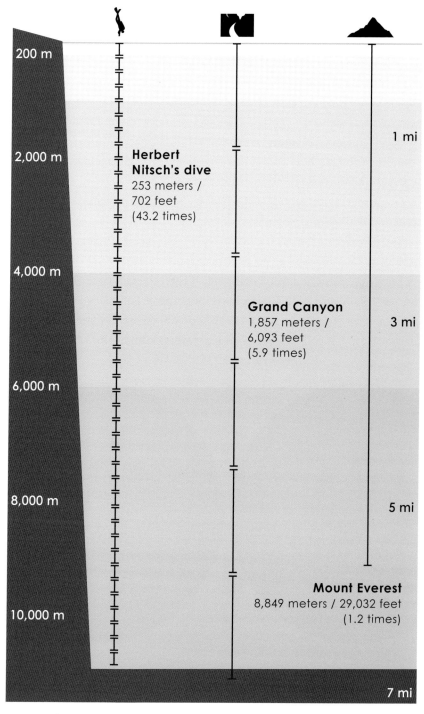

200 m

2,000 m

Herbert Nitsch's dive
253 meters /
702 feet
(43.2 times)

1 mi

4,000 m

Grand Canyon
1,857 meters /
6,093 feet
(5.9 times)

3 mi

6,000 m

8,000 m

5 mi

Mount Everest
8,849 meters / 29,032 feet
(1.2 times)

10,000 m

7 mi

A comparison chart showing how many times over you could fit Nitsch's dive, the Grand Canyon, and Mount Everest into Challenger Deep.

And if it were Grand Canyons all the way down, we'd need six of them to reach the bottom.

If we were lucky enough, and brave enough, to go all the way down (and if we survived), what would we see down there? What does the bottom of the ocean look like? It's not the flat, sandy landscape you might expect. The ground floor of the sea is a precipice-strewn hotbed of geologic activity. Mountains and canyons dwarf those found on dry land.

The rim of the Mariana Trench, the chasm where Challenger Deep is found, sits at about 6 kilometers (3.7 miles) below sea level.

Six kilometers is already exceedingly deep. In fact, only the most precipitous trenches and depressions plunge to such depths. If 6 kilometers of water were suddenly drained from Earth's oceans, this is all that would remain.

And if we were to stand at the edge of this new, meager Mariana Sea, Challenger Deep would still be 4,924 meters deep—nearly 5 kilometers, or 3 miles.

So, yes, Challenger Deep is deep. Exceedingly deep. But it is more than a geologic anomaly or curiosity to those fond of obscure facts and figures. Reaching Challenger Deep represents a depth of determination and spirit of exploration

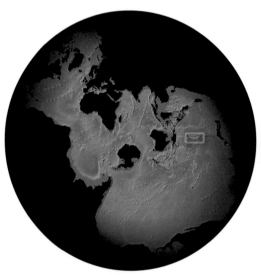
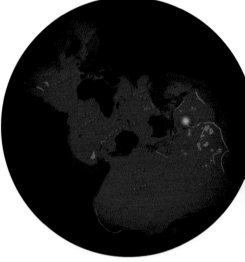

This map is rendered in the Spilhaus projected coordinate system, which effectively shows the global ocean as one interconnected system.

What would remain of the Earth's oceans if 6 kilometers of water were drained from them.

that should inspire us all. It bears witness to the will and capacity of science and engineering to build and pilot crafts that ferry humans to the remotest nook on our planet; it raises our hopes for discovering, mapping, understanding, and appreciating the rest of it—and beyond. Five years ago, less than 5 percent of the ocean was mapped to a high resolution. Today, almost 25 percent has been mapped—an incredible leap, but one that still leaves an astounding 75 percent of the ocean awaiting detailed mapping. For that, we have to rely on the courage, dedication, and ongoing work of scientists like Dawn Wright.

Getting there

" *Explorers from ages past, even from a few decades ago, would be dazzled by the precision with which it is now possible to determine where you are, whether on land, at sea, and even, most amazingly, under the sea.* —Sylvia Earle

By a historical accident, Challenger Deep is aptly named: it's hard to imagine a greater challenge (on this planet, anyway) than sending a person almost 7 miles under the ocean—and, crucially, back again. Victor Vescovo often remarks that, in deep-sea diving as in mountaineering, it doesn't count if you don't come back. Alive. Not only is the deep-sea pressure so immense that it can turn the human body into "pink slurry"—a grotesque but accurate description—the engineering needs to be fail-safe. You're on your own down there, in the absolute darkness and the hadal cold.

So let's take a moment to recognize the bravery of all who venture there, even in the 21st century, with advanced technology, materials, and engineering. In such extreme conditions, there's no margin for error, as the shocking implosion of OceanGate's *Titan* submersible reminded us all in June 2023.

Little wonder, then, that it took almost a century after HMS *Challenger*'s fortuitous sounding of the Mariana Trench for humans to descend to those depths. But even without any deep-sea dives, the *Challenger* expedition of 1872–1876 represented a major scientific undertaking for its time—a 200-foot former warship carrying a six-person scientific team and a crew of 241, retrofitted with state-of-the-art scientific equipment, a natural history laboratory, a chemical laboratory, and 181 miles of rope for dredging and sounding. It was a bold display of British imperial power, technological know-how, and scientific ambition.

In the course of an arduous three-and-a-half-year journey, covering almost 70,000 nautical miles, those onboard suffered hardships ranging from seasickness to smallpox to loss of life. Many crew members deserted; the German biologist died of a bacterial infection. The ones who stuck it out faced the gamut of dangers at sea—storms, sharks, cyclones, icebergs, you name it. And they dredged and dredged and dredged, in the process transforming our understanding of the deep.

A red sea serpent from Olaus Magnus's *Carta Marina*.

Until then, no one had known for sure what was down there. Early mapmakers had populated the blank ocean spaces on their maps with fanciful monsters, reflecting mariners' fears of the unknown. The 1539 *Carta Marina* by the Swedish geographer Olaus Magnus, for example, shows a giant red serpent that has a ship in a stranglehold; a giant amphibious owl being attacked by a snouty, spiky beast; and a sea hog, with a pig's head, curved horns, dragon's feet, a pair of eyes on each side, and a third pair (a spare?) on its belly. As subsequent science has shown, there are indeed remarkable creatures in the deep, but none quite like this.

In the Victorian age, a different theory took hold, banishing such monsters from the popular imagination—banishing, in fact, all life from the deep. The so-called azoic theory, championed by Professor Edward Forbes from the University of Edinburgh, maintained that *nothing* could live below 300 fathoms (about 550 meters, or 1,800 feet). According to Forbes's doctrine, based on some failed experiments at 230 fathoms, the abyss was a dead zone, a dark, impenetrable wasteland, devoid of life. Despite other explorers' evidence to the contrary, the azoic theory gained traction among scientists at the time. After all, they wondered, how could any form of life exist without oxygen, without food, and without light? Seemed like a no-brainer.

Charles Wyville Thomson and William Carpenter weren't so sure. In 1868, Thomson, a Scottish naturalist, and Carpenter, vice president of the Royal Society, embarked on an oceanographic expedition aboard HMS *Lightning*, seeking to dredge at greater depths. Their preliminary findings inspired them to return to sea the following year in HMS *Porcupine*. This time, when they dredged below 2,000 fathoms (about 3,657 meters, or 12,144 feet), they were rewarded with a bioluminescent bounty, a haul of glowing and sparkling life-forms, from worms to coral to brittle stars. "We concluded," wrote Thomson, "that probably in no part of the ocean were the conditions so altered by depth as to preclude the existence of animal life—that life had no bathymetrical limit. Still," he added, "we could not consider the question completely settled."

To settle the question, Thomson undertook another ambitious expedition, recruiting John Young Buchanan and John Murray as fellow members of the scientific team. (Carpenter, at 60, had bowed out.) Their goal: to survey the deep. Their vessel: HMS *Challenger.*

In addition to their extraordinary good fortune in sounding the Challenger Deep, the naturalists found what they were looking for: life, in the deep, wherever they sailed and wherever they dredged. An amazing abundance of it.

So much for the azoic theory.

With these findings, the *Challenger* expedition completely rewrote the scientific narrative of the day, opening up intriguing questions for future generations of ocean scientists to explore. What were these creatures? How did they survive down there? How could they even exist? Thomson himself marveled at "the recklessness of beauty," which produced such intricate, exquisite structures to live unseen—from a human-centric point of view—in the abyssal depths.

It would be another 54 years before a human set eyes on these creatures in their own undersea world. William Beebe was a self-taught naturalist who worked for the Bronx Zoo and, sponsored by wealthy donors in the Roaring Twenties, traveled the world, studying and collecting wildlife. Over time, his lifelong fascination with the ocean escalated into an ambitious research project: he wanted to survey and study all the life-forms within a defined area of the ocean, about 13 kilometers (8 miles) square and 3 kilometers (2 miles) deep. Britain's Prince George offered Beebe the use of a private island in Bermuda, other patrons provided a research ship, and, in 1928, Beebe set sail.

He soon realized that dredging wasn't going to work: he was netting only the mangled remains of those creatures unlucky enough to be caught. What else was down there? To understand the full scope of deep-sea life, Beebe decided, he had to go and see for himself. But how? Scuba diving didn't yet exist, and the helmet diving of the time was limited to depths of a few hundred meters.

Never publicity shy, and hoping to drum up support, Beebe announced his intentions in the *New York Times.* He was planning to build a small cylindrical vessel that would take him into the deep, becoming the first human to observe its creatures up close. When Owen Barton read this, alarm bells sounded. The young engineering student had already been working on plans for a submersible, hoping to be the first. Not only that, but he knew Beebe's cylindrical design would never work. The only shape that can withstand the pressure of the deep is a perfect sphere, since the otherwise crushing forces are distributed equally across its surface.

Eventually, Beebe and Barton formed a partnership, which culminated in the production of a clunky-looking hollow sphere, just under 1.5 meters (5 feet) in diameter, made of inch-thick cast steel and weighing over two tons on land. It had three windows made of fused quartz (the strongest type of glass then available), oxygen tanks for air supply, and a heavy entrance hatch that had to be bolted down before launch. It even had a telephone system on board.

Unlike today's submersibles, which are self-propelled, this contraption would be winched from a support ship, dangling from a steel cable all the way down. Beebe named it the *Bathysphere*, from the Greek word *bathys*, or deep.

In theory, it could work. The only way to find out was to … dive in it, and hope for the best.

Beebe and Barton did just that on June 6, 1930, when the *Bathysphere* was lowered into the Atlantic, off the coast of Bermuda, for its inaugural dive. The two men scrunched into the cramped space, trusting that all the engineering glitches had been worked out in test dives, but also knowing that the odds of a catastrophic leak or explosion were pretty high. Or, worse, that the cable that suspended them might fail, sending them plummeting to the seafloor in a sealed metal ball.

The dive was not without incident—water trickling in at one point, an explosion of sparks at another—but the *Bathysphere* made it to a depth of 245 meters (803 feet) before Beebe decided it was time to turn back. Climbing out of the hatch that day, they had made history. Beebe proudly announced: "The window to a wholly new world has been opened at last to human eyes." And it had.

Beebe had been the first human to observe how, as you descend deeper into the ocean, the red and yellow wavelengths of light disappear, filtered by the water, leaving only green, violet, and blue visible. (This explains why so many deep-sea creatures are red: a kind of invisibility cloak!) But what really moved Beebe, on that first dive, was the intense, otherworldly blue of the deep ocean, at the edge of darkness: "We were the first living men to look out at the strange illumination," he wrote. "And it was stranger than any imagination could have conceived. It was

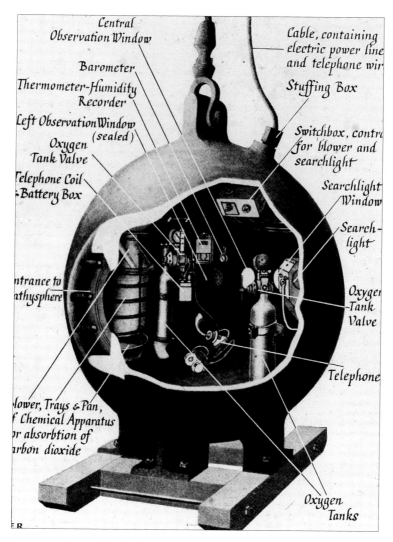

The labels on the diagram read:

Central Observation Window

Barometer

Thermometer-Humidity Recorder

Left Observation Window (sealed)

Oxygen Tank Valve

Telephone Coil Battery Box

ntrance to athysphere

lower, Trays & Pan, f Chemical Apparatus or absorbtion of arbon dioxide

Cable, containing electric power line and telephone wir

Stuffing Box

Switchbox, contro for blower and searchlight

Searchlight Window

Search-light

Oxyger Tank Valve

Telephone

Oxygen Tanks

The instruments aboard the *Bathysphere.*

of an indefinable translucent blue quite unlike anything I have ever seen in the upper world."

As early forerunners to Dawn and other women in ocean science, three team members were key to the mission's success: laboratory assistant Jocelyn Crane Griffin, technical officer Gloria Hollister Anable, and Else Bostelmann, an artist. Aboard the support ship, Griffin helped identify the creatures observed and collected on the dives. Anable maintained a phone connection between the bathysphere and the ship—its lifeline to the surface. And Bostelmann created stunning

Today, the *Bathysphere* sits on display at the National Geographic Museum in Washington, DC.

color illustrations, later published in *National Geographic Magazine*, of deep-sea life. Not only that, but Griffin and Anable also bravely descended in the bathysphere, to depths of 365 meters (1,200 feet).

Beebe and Barton themselves completed 33 more dives in the *Bathysphere* between 1930 and 1934, eventually descending to 922 meters (3,028 feet). They observed an eye-popping array of bioluminescent beings, and Beebe went on to write a best-selling account of his adventures. It wasn't entirely a happy ending, though: the scientific community expressed skepticism about some of Beebe's findings. And, as humans do, Beebe and Barton fell out. But their joint achievement could never be denied; they had indeed opened a whole new world to human eyes—one that had been there all along, the oldest and deepest on our planet.

> *When once it has been seen, it will remain forever the most vivid memory in life, solely because of its cosmic chill and isolation, the eternal and absolute darkness, and the indescribable beauty of its inhabitants.* —WILLIAM BEEBE

The challenge of Challenger Deep

In 1933, the *Bathysphere* was exhibited at the Chicago World's Fair, next to an aluminum sphere that had also completed a record-setting journey—but in the opposite direction. Attached to a hydrogen-filled balloon, this sphere, built by the Swiss physicist Auguste Piccard, had carried its inventor 15,781 meters (51,774 feet) into the stratosphere, another first for humanity.

It was a fateful juxtaposition—the *Bathysphere* and Piccard's sphere—because Piccard soon realized that the airtight, pressurized capsule he'd designed for balloon flight could be adapted for underwater exploration. And so he turned his attention to designing what he called the bathyscaphe (from the Greek words for *deep* and *ship*). With its small spherical cabin suspended beneath an enormous float, Piccard's bathyscaphe was, basically, the underwater equivalent of a hot-air balloon.

It might have been ungainly, by today's standards, but it worked. Piccard's brain wave had been to fill the float with gasoline, which is lighter than water and wouldn't compress. Additional air-filled ballast tanks would fill with seawater when vented, allowing the craft to descend; ditching about 10 tons of ballast weights was the low-tech way for it to ascend. Between 1948 and 1954, Piccard made a series of test dives, reaching a record-breaking depth of 3,150 meters (10,335 feet) in 1953. His companion on that dive was his son, Jacques, who would go on to make history of his own.

By 1960, Piccard's modified bathyscaphe, now named the *Trieste*, belonged to the US Navy, which had bought it in the spirit of Cold War competition. The space race was already under way, and who knew what advantages, military or otherwise, might also be gained by exploring the deep? While NASA set its sights on the heavens, the US Navy turned its gaze to the deepest known place on Earth: Challenger Deep.

On January 23, 1960, two pilots were bolted into the *Trieste* for its historic dive: Jacques Piccard, who had collaborated with his father on the design and testing of the bathyscaphes, and a 28-year-old US Navy submarine lieutenant named Don Walsh, who somehow thought risking his life at crushing depths in an experimental contraption might be fun. Apart from that nerve-wracking crack in the exit-hatch window at 9,012 meters (29,568 feet), the mission was a success, though their landing stirred up so much sediment that they were unable to see much. But they did glimpse something fishlike darting by: another nail in the coffin of the azoic theory.

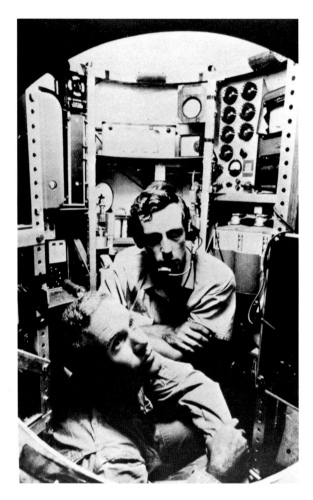

Don Walsh (*foreground*)
and Jacques Piccard
aboard the *Trieste*.

Some 52 years later, Don Walsh was on hand to advise the next human determined to reach Challenger Deep. Most of us know James Cameron as the director of blockbusters such as *Titanic* and *Avatar*, but there's a reason so many of his
movies have an aquatic theme: Cameron has been obsessed with the ocean all his
life. For *Titanic*, he made numerous dives to the famous wreck, contributing to
the development of deep-sea film and exploration technology. As a boy, inspired
by his passion for science and science fiction, he'd dreamed of traveling to the
ocean depths. And by 2012, in collaboration with National Geographic, he had the
wherewithal to turn that boyhood dream into a high-tech reality.

If the *Trieste* resembled a hot-air balloon, Cameron's submersible, the *Deep-
sea Challenger*, looked more like a long, green torpedo. (Inside, the capsule containing the pilot was a sphere.) The engineering team had spent seven years on

its innovative design, using technology and materials that simply didn't exist in *Trieste*'s day—materials such as syntactic foam, a high-strength but lightweight material made of hollow spheres bound together by a polymer. As a result, the *Deepsea Challenger* weighed significantly less than those early, lumbering bathyscaphes. It also had thrusters that enabled it to travel along the seafloor without stirring up an obscuring cloud of silt.

Despite all the tech and all the testing, Cameron's dive, on March 26, 2012, was glitchy. Some of the sub's systems malfunctioned, but Cameron arrived safely at the seafloor, with tools to collect scientific data, specimens, and images that Auguste Piccard could only have dreamed of. When he surfaced, with Don Walsh there to congratulate him, Cameron said in amazement: "In the space of one day, I've gone to another planet and come back."

This historic expedition—the first solo dive to Challenger Deep—helped shed new light on the secrets of the deep, bringing back 3D images and samples that led to the identification of at least 68 new species (mostly microbes). Yet, for a host of technical reasons, the *Deepsea Challenger* never dived again. It would be seven more years before another human-piloted sub reached the deepest place on Earth.

Chapter 4

THE FIVE DEEPS

> " *I once saw a map with all the unmapped areas of the ocean shown in black. I was shocked at how much of that map was black—most of the deep ocean was almost totally unmapped. How could this be?* —VICTOR VESCOVO

All Victor Vescovo needed to alter the face of ocean science was a thirst for exploration and the resources to make it happen.

Victor—let's refer to him informally here, as he has a starring role in Dawn's story—was born in Dallas, Texas, in 1966. From all accounts, he was the type of super-smart, sci-fi-obsessed kid who would make a space capsule out of a cardboard box, then soup it up with a power source and a functioning phone. He built a working radio when he was eight and a working computer when he was 13. And he was a voracious reader, who would read anything and everything, including the encyclopedia. Also, Victor says, he loved to look at atlases: "I don't know why, but I've always been fascinated by maps. As a kid, I could study them for hours. Sometimes, when I was little, I thought about becoming a geographer."

This early fascination with maps soon morphed into a passion for real-world exploration and adventure. After graduating from Stanford University, Victor began working in finance and, with his aptitude for math, eventually made a fortune in private equity. He also racked up graduate degrees from MIT and Harvard and served for decades as a reserve naval intelligence officer. But when he was 21, a spontaneous hike up Mount Kilimanjaro stoked his desire for even greater challenges. By 2011, he had become only the 38th person to complete the Explorer's Grand Slam: ascending the Seven Summits (the highest peaks on each continent) and skiing the North and South Poles.

Now, the quest for a new challenge consumed him. What next? What *hadn't* been done before—by anybody?

Victor Vescovo at the summit of Mount Everest, holding the state flag of Texas and the US flag.

Victor had been following the exploits of the British billionaire and explorer Richard Branson, who—in between planning missions to space—had announced a deep-sea initiative called the "Five Dives." Branson's ambitious goal, set in 2011, was to pilot a single-person submersible to the deepest point in each ocean.

From the start, Branson's project, branded as an "epic adventure," was beset by design challenges and technical failures. When the transparent dome on his prototype sub imploded under pressure testing, Branson called it a day. In 2014, the Five Dives project was quietly abandoned.

But Branson's quixotic mission had planted a seed in Victor's brain. Since he had already ascended to the highest points on Earth, why not also descend to its depths? It astonished him that, in the 21st century, humans had traveled to the moon and landed on Mars yet had only reached one of the five deeps (Challenger Deep). And nobody knew for sure where the other four were. Their current coordinates were best guesses, based on satellite data. Before he could even dive to the deepest depths, he would need to identify them.

Luckily for him (and for science), Victor had the financial means to pursue his new obsession. His first step was to establish Caladan Oceanic, a private company dedicated to advancing undersea technology and deep-sea expeditions. Like so many of the quirky names in the Victor-verse, the company's name comes from science fiction—in this case, Frank Herbert's *Dune*. Caladan's first mission would be the Five Deeps expedition, a deep-sea moon shot to reach unexplored parts of our own planet.

Victor's goal was to dive to the deepest known point in each ocean: the Puerto Rico Trench in the Atlantic, the South Sandwich Trench in the Southern Ocean, the Java Trench in the Indian Ocean, Challenger Deep in the Pacific, and the Molloy Hole in the Arctic. (The Molloy Hole isn't technically a trench because, at 5,550 meters/18,210 feet, it doesn't go all the way down to the hadal zone.) To achieve this, Caladan, starting from scratch, would have to outfit an expedition to travel around the globe, including both polar regions, and reach extreme ocean depths where no human had ever been.

First, though, Victor needed a sub. One that, ideally, would not implode in the abyss and would also be capable of repeated dives.

The tech

At the time, there weren't too many companies capable of building a submersible that could do what no submersible had done before. Victor's quest led him to Triton Submarines, a private submersible design and manufacturing firm cofounded by Patrick Lahey. Lahey accepted the challenge of leading the project, which eventually came to involve a complex international effort—and a giant headache. As Lahey described the yearslong process of trial and error, it was one of rigorous testing, modifying, and retesting, using each failure as a guide to success.

Vescovo's initial idea was simple: "At first, all I wanted was a craft that could safely dive to the bottom of the ocean and take some basic measurements and video—basically, a metal ball." An updated bathysphere? No way. Lahey had to persuade Victor that the sub should be a two-person machine, equipped with three viewports, instruments for scientific research, and a manipulator arm for retrieving samples from the seafloor.

Triton also insisted that, despite the expense, the sub be certified by an independent, third-party agency—for safety, of course, but also because many in the oceanographic community wouldn't set foot in a sub that wasn't certified. Almost a decade later, OceanGate's catastrophic loss of an uncertified sub confirmed the wisdom of this decision.

In the end, it took three years and some $37 million for Victor's state-of-the-art submersible to be designed, built, and tested, with some of the best minds in the business attacking the problem. Its production was a complicated global affair,

THE FIVE DEEPS

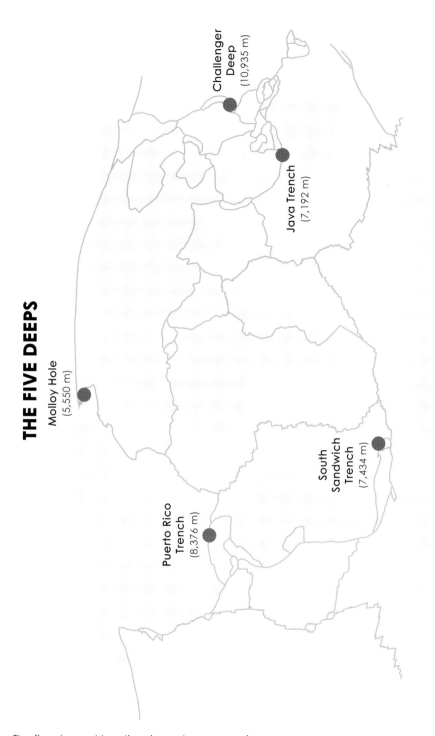

Challenger
Deep
(10,935 m)

Java Trench
(7,192 m)

Molloy Hole
(5,550 m)

South
Sandwich
Trench
(7,434 m)

Puerto Rico
Trench
(8,376 m)

The five deepest locations in each ocean region.

Patrick Lahey on the DSSV *Pressure Drop*.

with advanced components manufactured around the world and shipped to Triton's facility in Florida for assembly.

The final and most nerve-wracking step in this international odyssey was pressure-testing the inner titanium sphere that would contain the pilot and passenger. It was about 1.5 meters (5 feet) in diameter and weighed 8,000 pounds, and the only pressure-testing chamber in the world large enough to contain it was in St. Petersburg, Russia. This politically sensitive location added an extra dash of anxiety to the team's engineering jitters. Tension was running high. Would the Russian government intervene in some way? And would the 8.8-centimeter-thick (3.5-inch) sphere implode at pressures equivalent to 14,000 meters (45,920 feet), deeper even than Challenger Deep?

The pressure hull performed impeccably, validating Triton's design. The sub, at last, was good to go. Victor named it the *Limiting Factor*, after a spacecraft in Iain M. Banks's science-fiction Culture series. The name seemed a good fit because the sub had been designed to withstand the "limiting factors" of extreme ocean depth. The Culture novels were also the source of the weirdly cute names Victor bestowed on the three unpiloted landers built to support the *Limiting Factor* and carry out scientific tasks on the seabed: the *Skaff*, the *Flere*, and the *Closp*.

Meanwhile, Victor and Triton had been searching for a suitable support ship

The *Limiting Factor* at the South Sandwich Trench, February 2019.

to carry the *Limiting Factor* and its team around the world on a mission that would last at least a year and cover vast distances, including ice-bound regions. In the end, they ended up buying and refitting a vessel that had originally been commissioned by the US Navy—ironically, to hunt submarines during the Cold War.

The ship was not in the best of shape—to put it politely—but, as a naval hunting vessel, it was extremely quiet, critical for underwater acoustic communications. Meeting the requirements to "reclass" the vessel as a civilian research ship was a mammoth undertaking that went way over budget, but finally, in February 2018, the much overhauled vessel was certified as seaworthy. Among other upgrades, it had been fitted with sophisticated scientific labs and communications equipment (not to mention a fancy espresso machine for the crew). Victor renamed it the *Pressure Drop*, also from the Culture series, since it would "drop" the *Limiting Factor* into the pressures of the deep.

The team

The state-of-the-art equipment that Victor had commissioned—the submersible, landers, and support ship—made his mission possible, but the tech was by no means the whole story. These machines, advanced as they were, could not operate themselves. The success or failure of the Five Deeps expedition would depend, to a great extent, on the human factor—the team that Victor assembled.

The DSSV *Pressure Drop*.

Logistics and leadership would be critical to an expedition that involved confining dozens of people on a former naval vessel and sending them to some of the most desolate expanses of ocean for weeks at a time—all in a high-stress environment where life- and mission-threatening complications could occur at any time.

As the expedition leader, Rob McCallum, put it: "You take a ship that's been refitted from its original purpose and has never been used for this purpose, you're taking a submersible that's a prototype, fitting it with a sonar that's a prototype, fitting it with three landers that are all prototypes, to do something that's never been done before, with a group of people who come from 17 nations and have never worked together ..."

What could possibly go wrong?

A lot went wrong, especially in the beginning, as is inevitable in an undertaking of this scope and complexity. But, when it counted, everything went right. It took a large and talented team of people working under pressure cooker conditions to make that happen.

So let's meet some of the humans who made the Five Deeps and Victor's subsequent expeditions not only possible, but, in the end, a success.

The guide

How do you make what is not currently possible, possible? If it doesn't break the laws of physics, it can be done. —ROB McCALLUM

Rob McCallum has spent most of his life, in his words, "making complex things happen in remote places." He moved from his childhood home in Papua New Guinea to New Zealand to work as a ranger in that country's National Parks and then joined the United Nations as an adviser on the culture and landscape of the southwestern Pacific. This was followed by a spell with Deep Ocean Expeditions, at the time the only firm operating commercially in the deep ocean. There, he led dives to the wreck of the *Titanic*, among other exploits.

Eventually, he broke away and started his own company, EYOS Expeditions, an outfitter for people looking to push the limits of adventure at the planet's geographic margins. He takes pride in being able to respond positively when prospective clients walk in the door and say, "I'm told this may not be possible, but if it is, I hear you might be the one to do it."

Even by McCallum's standards, though, Victor's proposition posed a unique challenge. But McCallum's unflappable nature made him the perfect person to

Rob McCallum leads a morning briefing.

keep track of the expedition's countless moving parts. While at sea, his day would start as early as 3 a.m., with an analysis of the day's impending weather, and end as late as 11 at night, with the recovery of the last of the landers. In between, he was responsible for managing all the sub and lander launches and recoveries, as well as handling the logistics of the accompanying support boats, while also thinking and planning several days ahead.

McCallum's philosophy?

Between the starting line and the ultimate success, no matter how prepared you think you are, there are a thousand failures, so you've just got to work through them. ... You make failure a success. Let's have as many failures as we can today so that tomorrow we'll be in a much better position.

The captain

" *We had some really, really good people that knew what they were doing, and together, we got it done.* —STUART BUCKLE

A straight-talking, no-nonsense Scotsman, Master Mariner Stuart "Stu" Buckle followed his father's footsteps onto North Sea oil rigs and from there into officer training and captaincy. In 2012, he captained the support vessel for James Cameron's *Deepsea Challenger* dive, making him a natural for the Five Deeps.

But Buckle wasn't originally part of Caladan's team; at first, the *Pressure Drop* had been staffed with a yacht crew. Then McCallum, overseeing the preparations for the expedition, grew increasingly concerned about the crew's ability to spend months at sea while sustaining a serious scientific mission. He decided to reach out to Buckle, who, after checking out the ship, had a brief meeting with Victor.

Buckle recalls what may be the shortest job interview in history: "The ship was a mess, you know, it needed a lot of work. Then I had like a six-minute job interview with Victor in a coffee shop. I said to him, 'Look, I'm fairly sure I can do this, but I need to have complete control of the ship, and I need to be able to bring in the guys I need to get the job done.' And he just went, 'Sure,' finished his coffee, and walked out."

Captain Stuart Buckle at the helm of the DSSV *Pressure Drop*.

Buckle repaid Victor's faith by rising to meet the enormous responsibility placed on him. His first task was overseeing the arduous process of refitting the *Pressure Drop* from its neglected state in dry dock—something that wouldn't have been possible without Buckle's supervision, Victor says.

During the missions themselves, Buckle was accountable for every single aspect of the *Pressure Drop*'s operation, no matter how minute, from ensuring ample food on board to positioning the ship in precisely the right spot for the *Limiting Factor*'s launches. And if you think that might be plain sailing, Dawn says, "Imagine that you are a tiny drone and that you're trying to find an individual seat in a gigantic football stadium. You are in complete darkness and high winds, and you have only the drone's tiny camera and light to guide you to the exact seat."

Buckle also personally recruited the 36-person crew, drawing on his years of experience to assemble the right mix of people and skills. This made him feel all the more responsible for the success of the mission. After all, he said, "Most of the guys I asked to come here left good, secure, well-paid jobs to come and work [under] me, for a guy they'd never heard of. ... I pinned my whole professional reputation on this job and being the guy to get it done."

No pressure!

The mapper

 I have to explain it to people. The ocean is not mapped!
—CASSIE BONGIOVANNI

Sailing around the world to map uncharted depths was not something Cassie Bongiovanni dreamed she'd be doing just months after completing her master's degree in ocean mapping at the University of New Hampshire. She was newly on the job market when she received a cryptic email, forwarded by a contact from NOAA—something vague, she recalls, about "120 days at sea to map some stuff." She sent off her résumé and soon received a call from a man with a New Zealand accent. It was Rob McCallum, inviting her to join the Five Deeps expedition as lead mapper.

For someone so passionate about mapping the ocean, only about 20 percent of which had then been mapped, it was the opportunity of a lifetime. The expedition would take Bongiovanni to those blank spots on the map that she longed to chart. "I couldn't sleep that night," she recalled. "I was so excited about the possibilities."

Cassie Bongiovanni at her
workstation aboard the
Pressure Drop.

But, as a 25-year-old woman joining a more experienced, almost entirely male team for long stretches at sea, she had to wonder how things would work out. Would her expertise be respected? And was this long-shot mission even going to succeed? It could be a glorious adventure—or a dismal failure.

Nonetheless, in December 2018, Bongiovanni reported for duty in Curaçao, where the state-of-the-art EM 124 multibeam sonar was being installed on the *Pressure Drop.*

Because sonar can be running at any time, mapping effectively becomes a 24-hour operation. Typically, this is a job done by a team of three or four mappers, alternating day and night shifts. But during the first phase of the expedition, Bongiovanni worked alone, which meant that, in addition to monitoring the daytime sonar output, she also had to review the overnight returns for gaps in the data. While working to pinpoint the deepest spot in the Puerto Rico Trench, she barely slept. After that, Victor hired another mapper, Aileen Bohan, who shared the round-the-clock workload for the rest of the trip.

Bongiovanni is justifiably proud of the fact that, over the course of 10 months, the Five Deeps expedition mapped about 550,000 square kilometers (212,000 square miles) of seafloor—an area half the size of Brazil. She's equally proud that she convinced Victor to donate all the maps from the Five Deeps expedition to Seabed 2030, a project Dawn also champions. But what really gratified her was getting Victor to share her excitement and enthusiasm about the maps they were producing. In the South Sandwich Trench, then almost entirely uncharted, Bongiovanni produced sharp three-dimensional images of what had previously been blurry satellite predictions:

> Seeing that for the first time, he was *ecstatic*, saying, 'This is 100 percent worth the millions of dollars I paid to get this thing [the multibeam sonar] put on here!'

He printed off pictures, posted them all over the ship—he was like, "Look, this is what was there, and this is what we have now! This is what we're doing!' He was so pumped about it. ... It became a pretty good source of pride for everybody on the ship.

The scientist

 Challenger Deep is one of those places that we need to understand so that we can really understand how our planet works, from top to bottom.
—DAWN WRIGHT

Who better to oversee a mission to explore the deepest reaches of the oceans than the person who literally wrote the book on the hadal zone? That would be Dr. Alan Jamieson, author of *The Hadal Zone: Life in the Deepest Oceans* (Cambridge University Press, 2015), who was persuaded to join the expedition as leader of the scientific team.

Unlike Dawn, Jamieson wasn't initially drawn to ocean science by a passion for the deep. Rather, his academic background was in industrial design, and he stumbled into applying that to underwater vehicles because he was a student in Aberdeen, Scotland, a city that subsists largely on the offshore oil and gas industry.

After a series of odd jobs, he happened upon an opening as a low-level technician in a lab run by a professor who was researching deep-sea fish. The prof helped guide Jamieson through a master's program that soon evolved into a PhD. Jamieson's original goal was to build landers capable of operating consistently in the hadal zone, but, as he explained, "I started learning more and more about biology and ecology and fish behavior, as that seemed inextricably linked to the success of the gear I was building."

The hadal landers succeeded, and Jamieson became one of the preeminent deep-sea scientists in the world. He wrote his book on the hadal zone and took a post as a lecturer at Newcastle University but remained very much a field researcher. His expeditions to far-flung, geographically diverse locales grew "crazier and crazier," until he reached a point where he felt he'd hit a wall in terms of what he could do with his hadal exploration technology.

That made it perfect timing when Patrick Lahey reached out, saying that he was looking for a lead scientist on a mission to circumnavigate the planet while

A new species, believed to be a sea squirt or stalked ascidian, that was discovered in the Java Trench.

diving some of its deepest trenches, most of which had never been dived before. There was no hesitation on Jamieson's part: "This was my kind of crazy," he said.

Jamieson designed Caladan's three scientific landers, the *Skaff*, the *Flere*, and the *Closp*, capable of full-ocean depth. And, once aboard the *Pressure Drop*, through all the inevitable frustrations and disappointments of conducting research under such extreme conditions, he made sure that the science got done. "Without Dr. Jamieson," said Victor, "we wouldn't have executed a tenth of the scientific research we did. He, more than anyone, made the [expeditions] a scientific—not just a technological—success."

Jamieson's scientific accomplishments, during the Five Deeps and subsequent expeditions, can be summed up in some impressive numbers:

- Over 100 deep ocean deployments of the three landers
- Over a million kilometers (621,371 miles) of water column data captured
- 400,000 biological specimens collected
- 30-plus suspected new species discovered
- Caught on video (via the landers) the deepest fish, octopus, and jellyfish ever recorded

The expedition

 Who says there is nothing left to explore on this planet? There is plenty to explore, and learn, in the oceans. I believe that the last great, true exploration left on planet Earth is to map the entirety of our seafloor. —VICTOR VESCOVO

In December 2018, after extended sea trials in the Bahamas, the *Limiting Factor* headed for its first deep ocean dive, the Puerto Rico Trench in the Atlantic Ocean. Even after three years of planning and testing, no one could be 100 percent sure that the sub and the landers were ready for prime time—the sea trials had been plagued by failures and mishaps, which the team had been working frantically to fix.

The Puerto Rico Trench, the second deepest on the expedition's itinerary, was formed millions of years ago by the North American Plate pushing slowly under and up against the Caribbean Plate. Cassie Bongiovanni, the mapper, had some incomplete data to work with, indicating that the trench was about 8 kilometers (5 miles) deep, but she needed to definitively establish the deepest point for Victor and the *Limiting Factor* to dive. Over a sleepless couple of days, Bongiovanni surveyed an area roughly the size of Rhode Island so that she and team geologist Heather Stewart could identify the exact spot, later confirmed at 8,376 meters (27,480 feet).

While Bongiovanni and the sonar equipment were working overtime, disaster struck the sub. Victor and Lahey were on their third and final test dive, to 1,000 meters (3,280 feet), when Victor decided to try out the manipulator arm. Lahey describes what happened next:

> We're on that dive, you know, we get to the bottom, he deploys the arm, and the arm sort of unceremoniously falls off. The alarms panel is lit up like a Christmas tree—there are alarms for just about everything you can have an alarm for—and at that moment I know, I *know*, Victor was convinced we were done. *We were done.*

Victor was, indeed, discouraged, as were scientists Jamieson and Stewart, who'd planned to collect samples with the $350,000 part that had just plopped onto the seafloor, never to be retrieved. As Lahey had feared, Victor was ready to walk, write the whole thing off, wondering aloud if there were major design flaws that could never be fixed. But Lahey somehow persuaded him to give the team 36 hours to attack the problem.

Triton's principal design engineer, John Ramsay, and electrical engineer, Tom Blades, worked around the clock to determine what was up with the *Limiting Factor*. As it turned out, losing the manipulator arm was a stroke of luck, in engineering terms. It allowed the team to completely rewire the sub and fix the electrical problems that had been plaguing the test dives.

The following day, December 19, 2018, Victor made his historic descent to the bottom of the Puerto Rico Trench without a hitch. It took the *Limiting Factor* two and a half hours to reach the bottom and two and a half hours to ascend, with 45 minutes spent exploring the seafloor. To Victor, the sandy bottom looked like the surface of the moon—except for the cracked oil drum embedded there, an unwelcome reminder that the ocean has long been used as a dumping ground.

All the same, champagne corks popped on the *Pressure Drop* that night. Not only had Victor achieved his goal of being the first person to reach the deepest point in the Atlantic Ocean, but the *Limiting Factor*'s cameras had recorded at least four new species at full ocean depth. It was an exhilarating turnaround for the mission.

The next stop was the South Sandwich Trench, in the Southern Ocean, which surrounds the frozen continent of Antarctica. The waters here are notoriously hazardous, the coldest and roughest on the planet, littered with icebergs. As the team waited for a window of favorable weather, the harsh conditions—and the seasickness—were beginning to take a toll on their morale.

Once again, the mappers (Bongiovanni and Bohan) had to work nonstop, on a pitching and heaving ship, to determine the deepest point in the 965-kilometer-long trench (600 miles), which was almost entirely unmapped. When they found the exact spot, at 7,344.6 meters (24,388 feet) deep, its location was a surprise to everyone.

During Victor's solo dive, the communications system on the sub failed, spooking his support team, but he continued his descent, delighted by the abundance of marine life he observed on the way down—another surprise in these cold latitudes. This was, he realized, a glimpse into a true ocean wilderness, an underwater ecosystem that hadn't yet been disrupted by pollution or overfishing. As he would discover, there weren't many of those left on planet Earth.

But the mission's main scientific contribution in the Southern Ocean was the mapping. Over the next several days, the *Pressure Drop* sailed the entire length of the trench, taking accurate readings for the first time. As the bathymetry appeared on their screens, Bongiovanni and Bohan felt like astronauts on a new

Dr. Don Walsh (*left*) congratulates Victor (*right*) after his successful solo dive, as Dr. Patricia Fryer (*center*) looks on.

planet, seeing massive geologic features that no one had ever seen before. By the end, as Bongiovanni noted with pride, they'd collected the first complete dataset of the South Sandwich Trench.

After a couple of nail-biting but successful dives, the team had hit its stride. Unstoppable now, the expedition sailed on to dive the Java Trench in the Indian Ocean, with a new manipulator arm installed (much to Jamieson's relief), and finally the Molloy Hole, in the Arctic. Along the way, there were detours to dive and map other significant trenches and fractures, and even to visit the wreck of the *Titanic*.

But the dive that added Victor's name to an elite group of undersea explorers and that would eventually allow Dawn to fulfill her lifelong dream took place on April 28, 2019—into Challenger Deep.

The deepest dive

On board the *Pressure Drop* that day was an energetic and enthusiastic 88-year-old: Don Walsh, whom Victor had invited to witness his historic dive, 59 years after Walsh's own. Also present was Dr. Patricia Fryer, from the University of Hawaii, a marine geologist (like Dawn) who'd spent 30 years studying the Mariana Trench and had, in fact, discovered its second-deepest point, the Sirena Deep. For Fryer, this represented a rare opportunity to study rock or sediment samples that the sub's manipulator arm might be able to retrieve from the deep. All three of the expedition's landers would also be deployed, in this culminating dive for science, for glory, and for the record books.

When Victor successfully completed his solo dive into the Eastern Pool, Don Walsh was the first to shake his hand, marveling at Victor's mettle. Three days later, Victor made history again with a repeat dive, to a record depth of 10,935 meters (35,876 feet). The third dive was a salvage mission—the deepest ever—to free the lander *Skaff*, which had become stuck in the mud on the bottom.

After a fourth dive, piloted by Lahey, Victor and Jamieson made a science dive into the Sirena Deep, spending three hours on the seafloor and capturing imagery that Jamieson hoped would help scientists understand life at full ocean depth—organic compounds that don't depend on light, as life does on Earth, but derive their energy from chemosynthesis. As Victor reflected afterward, "Reaching Challenger Deep and Sirena Deep is *really* hard. Thousands have climbed Mount Everest, hundreds have been into space, but only about 30 people have been to the bottom of the ocean. Surviving a trip to explore down there is exhilarating, and there are amazing things to see."

Some months later, an email would land in Dawn's inbox, inviting her to see those amazing things for herself.

Chapter 5

NO LIMITS

> *We really should know what's on our own planet, and we should dive all these trenches to find out if they're similar or different. And if you think that it's just going to be boring and we're not going to find anything new, have you ever studied the history of science? You don't know what you're going to find, and it could be extraordinary. So, let's get at it!* —VICTOR VESCOVO

Having achieved his dream of diving the Five Deeps, Victor hung up his blue jumpsuit, plopped down on the couch, and had a cup of tea. ... Well, not exactly. Not even close.

Victor was nowhere near finished with deep-sea exploration. He'd won his laurels, certainly, but he still had the equipment and the team, so he was determined to keep going.

In 2020, he decided to launch a series of new expeditions. First, a dive off the southern coast of France to the *Minerve*, a French submarine that had mysteriously disappeared in 1968 and had only recently been located. Accompanied by the captain's son, Victor laid a memorial plaque at the wreck, honoring the 52 lives lost. Next, the *Limiting Factor* would make a series of first-time dives in the Mediterranean, the Red Sea, and the Indian Ocean.

For Cassie Bongiovanni, though, the most exciting leg of the expedition would be the Ring of Fire: a zone along the rim of the Pacific that is a cauldron of undersea volcanic and seismic activity, generating most of the earthquakes and volcanic eruptions on Earth. Bongiovanni's team would eventually map eight trenches in this zone, including, for the first time, the Philippine, Yap, Palau, Kuril-Kamchatka, and Aleutian Trenches.

During the trip, Bongiovanni also reached a personal milestone: 1 million

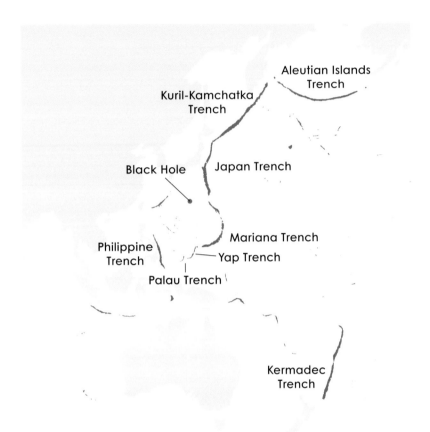

The trenches Caladan Oceanic dived to in the Pacific Ring of Fire.

square kilometers (386,102 square miles) mapped at sea. She and her team would go on to map more than 400,000 square miles of seafloor, data that would be donated to the Seabed 2030 project. An impressive achievement for anyone, but especially for an ocean mapper just two years out of grad school.

But the news that got Dawn's attention, at home in Redlands, California, was that the expedition would first be returning to Challenger Deep—to make a series of repeat dives and to collect heaps of additional data. The goal this time was not just to dive Challenger Deep but to survey all three of its pools and to map it intensively.

That's when Victor first approached Dawn—not, at first, to dive, but to support this ambitious mapping project. And, wonderfully, to support the dive of a fellow scientist, another super-accomplished woman, and a longtime friend of Dawn's: Dr. Kathryn Sullivan. At the time, Kathy Sullivan had already made

Dawn (*left*) with Kathy Sullivan (*right*) and former Esri Ocean Industry Manager Drew Stephens (*center*) at the Esri Ocean Forum, 2017.

history as the first US woman to walk in space. She would go on to make history as the only person to walk in space *and* visit the deepest place on Earth.

The "most vertical person in the world"

> *Get hooked on science—in whatever area you choose—and you're off on a grand adventure, exploring life with hands, mind, and heart.*
> —Kathy Sullivan

Trained as an oceanographer, Sullivan set her sights on the stratosphere as soon as flying there was an option for women. In 1978, she was selected as one of the first women to train as an astronaut for NASA. The competition was fierce that year—over 8,000 applicants for 35 slots—but six women were accepted into NASA's historic class, among them Sullivan and Sally Ride, the first US woman to fly in space. As if that weren't remarkable enough, Sullivan was only 26 at the time and already a PhD in geology. It was the lift-off to a brilliant career.

NASA portrait of astronaut Dr. Kathryn Sullivan.

First, though, a question: Why would an oceanographer aspire to be an astronaut? Sullivan's answer would probably be, "Why not?" Driven, since childhood, by intense curiosity and "a lust for adventure and travel and exploration," Sullivan says she applied to NASA partly to deepen her understanding of the Earth—to see our planet from orbit with her own eyes. Also, clearly, she loves a challenge. And, she says, she had plenty of experience planning and managing scientific expeditions, "high-stakes, no-kidding, real-world practical experience" that would transfer from the deep ocean to deep space.

"I loved the expeditionary part of oceanography best of all," she explains, "being out at sea, adapting to what odd circumstances came your way, bad weather or broken equipment. I adored that." That ability to adapt, to change course, and to follow an internal compass has taken Sullivan's career to the heights, as an astronaut, and to the depths, as an ocean scientist. In Dawn's words, "She's done it all now," earning her the title "the most vertical person in the world."

Sullivan first blasted off into space, in the space shuttle *Challenger*, on October 5, 1984. (*Challenger*, by the way, was named after the British naval research vessel HMS *Challenger*, which gave Challenger Deep its name.) Then, on October 11, she made history when she became the first US woman to step out of a spacecraft and "walk"—i.e., float in the void of outer space while tethered to a shuttle orbiting at 17,500 miles per hour. Her assignment that day was to test an orbital refueling system.

There's a story that Sullivan's father (an aerospace engineer) loves to tell about that walk. He was at mission control that day, in the viewing room, to witness the historic event. From where he stood, he could see the flight surgeon's console, which was monitoring Sullivan's vital signs. As she stepped out of the hatch, her heart rate spiked to 78—for just a moment, before settling back down to 60, where it remained for the next three and a half hours as she coolly completed her task. "I was just ready to go," Sullivan said. "I felt completely calm."

Sullivan served as an (exceptionally cool-headed) crew member on two more space missions, including the 1990 launch of the Hubble Space Telescope, the subject of her book, *Handprints on Hubble*. But when she left NASA in 1993, she found that the door was still open to her as an oceanographer. Accepting a different kind of challenge, "a different kind of jigsaw puzzle," she took on the role of administrator, serving as chief scientist of NOAA and later as NOAA's director under President Barack Obama.

There are a couple of pithy sayings that Sullivan likes to repeat, insights she's gleaned from both crewing a spaceship and running a massive government bureaucracy. One is a line from President Dwight D. Eisenhower on his experience as general: *"Plans are nothing, but planning is everything."* Another favorite: *"Stay smart and stick together."* If you think about it, these words are excellent advice for surviving on a space shuttle, being an effective leader, or piloting one's own career. And also, as it happens, for making a successful dive to the depths of the Pacific at the height of a global pandemic.

The "Magic School Bus"

Out of the blue, Sullivan had received an email from Victor Vescovo, inviting her to dive with him into Challenger Deep. As a supporter of greater diversity

Kathy Sullivan on her 2020 dive to Challenger Deep.

in science and technology, Victor believed that it was time for a woman to get down there—and not just any woman. "Dr. Sullivan has been a great explorer her whole life," he said. "I couldn't think of anyone better to be the first woman to the bottom of Challenger Deep." Of course, Sullivan jumped at the chance.

There was only one slight problem: a tiny virus known as COVID-19. For an entity one-hundredth the size of a grain of sand, COVID-19 was able to wreak enormous havoc across the globe in just a few months. Unluckily for Dawn, early 2020 marked the onset of the pandemic, which drove entire communities into lockdown as the death count rose, upending societies, economies, and even the best-laid plans.

Dawn picks up the story from here, recalling how, in April 2020, she and her colleagues from Esri were invited to support Sullivan's dive. Dawn was eager to be there for Sullivan, who was not only a respected peer but a friend. But the timing was terrible. "We got the request from Victor," she says, "not to dive, but to support the bathymetric mapping program of the ship over the Challenger Deep area, and to be there to witness Kathy Sullivan's dive and to celebrate with her onboard. But the request came in April 2020, just as we were beginning lockdown due to the pandemic. And they were planning to dive in June 2020, which was going to be impossible for anyone at Esri to take the risk of travel." Although Dawn was disappointed (to put it mildly), the connection between Dawn and Victor and Esri had been made, which eventually led to Dawn's own dive.

Victor, however, wasn't going to let something like a global pandemic stand in his way, and so the expedition went ahead as scheduled. On June 7, 2020, he piloted the *Limiting Factor* to the bottom of Challenger Deep, with Sullivan on board, making her the first woman to descend there (Dawn would be the fifth).

A highlight of Sullivan's dive was the call she made afterward from the DSSV *Pressure Drop* to the NASA astronauts on the International Space Station (ISS). The astronauts congratulated her, together they celebrated the technology that had carried them to the seafloor and to the stars, and Sullivan later said: "As a hybrid oceanographer and astronaut, this was an extraordinary day, a once-in-a-lifetime day, seeing the moonscape of the Challenger Deep and then comparing notes with my colleagues on the ISS about our remarkable reusable inner-space, outer-space vehicles."

As the first person to have experienced them, how does Sullivan compare these two extreme adventures? According to her, the trips couldn't be more different:

> The process of getting into orbit is intense and explosive, and short. You're riding a bomb, you're embedded in this crazy ball of energy, and in 8 1/2 minutes, you know, you're in a wildly different place.
>
> You get into orbit, you're a couple hundred miles above the Earth—any direction you go you can see probably nearly 1,000 miles. You get 16 sunrises and 16 sunsets in every 24-hour day. You're just stunned by the number of stars that you can see, so that there isn't really an inky black of space that's anything like the inky black of the deep sea.
>
> And conversely, going deep into the sea is a smooth, serene, very calm elevator ride. It took us four hours to go from the surface to nearly 36,000 feet. ... When you get to your destination, in the one case you have *no* air pressure outside the window, and the other case you have 16,000 pounds per square inch, 8 tons per square inch. You can see probably about 1,000 miles in any direction when you look out the viewport of a spacecraft and then maybe 30 feet when you look out the viewport of a submarine. In orbit, you're going from the daylight side to the night side, you know, every 45 minutes, so from pizza-oven hot to Antarctic cold and back every 45 minutes. Of course, in the deep sea, there's just this continual cold soak of the essentially freezing water temperatures through the metal of your sphere.

Sullivan also often marvels at what she calls the "Magic School Bus" effect of being transported to these otherworldly environments: "You're in a craft that's taking you somewhere that you otherwise cannot possibly be and have no business being," she says. "It's, of course, not magic, it's good engineering and science. But the experience has always seemed to me to have a dimension of magic to it."

 I've seen things no other human being has, and they've absolutely changed me in a lot of ways. ... We are all sharing this one home, this one little spaceship Earth. We've got to look at it as a place that you cherish and a place you want to protect. We're on it together—and we're intimately connected. —KATHY SULLIVAN

More firsts

Another extraordinary woman was the next to dive: Vanessa O'Brien, an explorer and adventurer with a record-breaking list of accomplishments. To name just a few:

- She holds a Guinness World Record for climbing the highest peak on every continent, the Seven Summits, in 295 days.
- She completed the Explorer's Grand Slam in just 11 months.
- As a dual national, she was the first British and first American woman to climb K2, the second-highest (but most challenging) peak on Earth.
- She rode with racing legend Mario Andretti at speeds of 322 kilometers (200 miles) per hour.

Oh, and she also allowed herself to be shot at point-blank range to test bullet-proof vests. ... Apparently, there's no challenge that O'Brien will not accept!

On June 12, 2020, she added a dive to Challenger Deep to this formidable list, making her the first woman to have climbed to Earth's highest point (Mount Everest) and dived to its lowest.

Just over a week later, history was made again when Victor piloted a very special passenger to Challenger Deep: Kelly Walsh, son of the *Trieste* captain, Don Walsh, who'd made the first descent in 1960. Kelly Walsh, arriving at the same spot on the ocean floor 60 years later, described the experience as "a hugely emotional journey" for him.

"I have been immersed in the story of Dad's dive since I was born," he said after his own dive, noting that "the leap in technology from 1960 is immense. Dad spent 20 minutes on the bottom and could see very little. I spent four hours on the bottom with excellent lighting and a 4K camera running the whole time. We had complete control over our vehicle, great lighting, manoeuvrability, and a comfortable cabin, whereas Dad had none of those things."

For his part, Don Walsh was proud and delighted to see his son follow in ... well, not his footsteps, but in his wake, to the depths of the Western Pool. It was a unique father-son bond, an experience shared by only a handful of people on the planet.

Next up, on June 22, was Dr. Ying-Tsong Lin, a deep-water acoustics expert from the Woods Hole Oceanographic Institution, who became the first person of Asian descent to visit Challenger Deep (and whose equipment, aboard the *Limiting*

Factor, picked up the reverberations of an earthquake thousands of kilometers away). Afterward, in awe, he wrote:

> The sub *Limiting Factor* is a space-time capsule bringing us to another world, which has not been touched for millions of years. Looking at the sand waves on the bottom of the world, thinking how long it took for the weak currents at that depth to build them up, space and time just collapsed; I was watching a million years of evolution in just an instant.

"Going home"

Challenger Deep lies within the territorial waters of the Federated States of Micronesia (FSM), but it wasn't until March 2021 that a Micronesian scientist dived in the sub to see it for herself. That scientist was Nicole Yamase, a marine botanist and, at the time, PhD student at the University of Hawaii. Victor had invited her because, he acknowledged, it was high time for someone from the FSM to visit the trench, which is right in their backyard.

Growing up on the islands of Micronesia, surrounded by the waters of the Pacific, Yamase (like Dawn) felt profoundly connected to the ocean as a child. "We were always surrounded by water," she says. "We'd go snorkeling with my dad and he'd point out the coral, fish, and algae. My dream to go into marine biology stems from those experiences." Currently, driven by a sense of urgency as the planet heats up, she studies the effects of climate change on macroalgae and nearshore marine plants.

For Yamase, science represents a way to complement and deepen the traditional knowledge of the Pacific Islanders, who have observed the ocean for centuries. "Our ancestors were scientists from the very beginning," she points out. "They observed and collected data ... they tested and tried new things." At the same time, she hopes that her path as a marine scientist will inspire more women from the Pacific Islands to pursue higher education and careers in STEM. Traditional culture may have kept women from venturing into the ocean, but, Yamase says, they belong there: "We belong all the way at the bottom, and everywhere in between too."

When she climbed into the *Limiting Factor* on March 11, 2021, Yamase—like all who dived to Challenger Deep—brought some personal items to take down with her. Among them were the FSM flag, a traditional *mwaramwar* necklace made of cowrie shells, and photos of two botanists who had inspired her, Dr. Isabella

Nicole Yamase aboard the DSSV *Pressure Drop* at the time of her dive.

Abbott and Dr. Roy Tsuda. Most meaningful of all was a small model wooden canoe, which belonged to her father and which, to Yamase, represented her family, the Pacific Island community, and her oceangoing heritage.

When Yamase finally arrived at her destination that day, her first thought was for her community and her ancestry. "I couldn't believe my eyes when I saw the fine silt bottom of the Challenger Deep through the small window," she recalls. "We were hovering two meters off the ground. This was the moment I was preparing for, and it was finally here. All I could think about was how proud my ancestors and the whole Pacific Island community would be."

In many cultures—and for many people—the deep ocean is feared as a dark, scary, and dangerous place, the domain of monsters. "But to Pacific Islanders," Yamase says, "it's the opposite. This is where magic is. This is where life was formed. This is where our islands were pulled out of. This is what our legends talk about." For her, a descent to the deepest spot in the Pacific felt almost like a homecoming: "It was almost spiritual," she says. "I felt like I was going home. ... I felt so at peace."

That could have been the end of the story, with Yamase flying back to Hawaii and resuming her doctoral studies. But, a year later, she was back aboard the

Pressure Drop, this time to support Caladan's 2022 expedition—the expedition of Dawn's dive and, as it turned out, the last such expedition for Victor and his crew.

A "pipe dream"

In May 2021, an email landed in Dawn's inbox from Victor Vescovo. Nothing unusual about that: she and Victor had stayed in touch and had, in fact, worked with a team from Esri to create a series of ArcGIS StoryMaps stories ("The Deep"), covering Caladan's 2021 expeditions in, well, depth.

But this email stopped Dawn in her tracks.

It read, in part:

> I am doing some long-range planning for next spring when the ship will be back at the Mariana Trench and wanted to know if you might want to do a dive there with me? Specifically, Challenger Deep. My happy place.

> Everything is very up in the air right now, but I think we might be back there in March, plus or minus a month. Just wanted to see if you might have the flexibility to come out to Guam for 7–10 days to do a dive.

> I remain a huge supporter of women in STEM, and my team remains a big user of Esri's tools, and I think it would be great to have you go down and see the place for yourself.

> Feel free to think about it, of course.

Think about it? Was he kidding? Asked how she felt when she received this invitation, Dawn says, "I was actually quite numb." After the disappointment of the Kathy Sullivan trip, which Dawn couldn't join, she hadn't dared hope for another chance. And so, Dawn says, "I was numb, but I was ecstatic."

As a marine geologist who studies how the seafloor is formed, Dawn had dreamed of this opportunity for the past 20 years—an opportunity to see the deepest, least-known, and most iconic spot on the ocean floor. For her, she says, visiting Challenger Deep had been "a pipe dream," one that she never thought would come to pass.

And there was another reason to celebrate, equally important. "Victor has asked me to do this dive in order to be the first Black person to ever visit Challenger Deep," Dawn said at the time. "So I'm taking that very seriously. I'm hoping that this will inspire and encourage little children, young people, certainly young people who are Black and who are not seeing ocean exploration and ocean science as something that they can do, or something that is welcoming to them. I want to turn that idea completely on its head and to say, Please consider doing this, we need as many different kinds of people as possible exploring the ocean."

A few weeks later, Dawn and Victor met on Zoom to discuss plans for the trip, still about a year away. Meanwhile, Dawn was gathering a team of seafloor mapping experts from Esri and Esri's partners to brainstorm with Victor on the technical aspects of the dive—how to handle data from the portable sidescan sonar system that would be used for the first time at full ocean depth. Dawn would serve as mission specialist and sonar operator, testing a prototype system built specifically for Caladan by Deep Ocean Search of France and Mauritius.

With this equipment, Dawn's goal was to make a backscatter image of an unvisited area in the Western Pool, where there had been far fewer dives than the Eastern Pool.

Dawn explains that sonar systems typically collect two types of 3D data: seafloor depth and backscatter. Bathymetry measures ocean depth, but technically speaking, it's a measure of the travel time of sound in water—i.e., how long it takes for a sound pulse to get to the bottom of the ocean and bounce back up to your instrument, which then translates to a depth. For context, it takes a voice signal about seven seconds to reach the bottom of Challenger Deep.

"And then," Dawn adds, "we have all kinds of sophisticated algorithms for making sure that we're getting the proper time because the speed of sound changes according to the temperature and the salinity of the water, which varies throughout the oceans."

Backscatter, on the other hand, is a measure of the *strength* of sound reflected off the seafloor and received by the sonar system. Different types of seafloor scatter sound energy differently, telling us, indirectly, what they are made of. Harder surfaces on the bottom, like rock, reflect more sound than softer ones, producing a darker image, while bumpier surfaces, like coral reef, return a more complex signal than smoother ones. With this kind of data, oceanographers and data scientists can use mapping software to create detailed 3D maps of the seafloor.

Technicalities aside, Dawn's findings would add another puzzle piece to our growing picture of Challenger Deep, offering greater insight into its geology and formation. With this goal in mind, Dawn and the Caladan team came up with a dive plan for July 2022: the *Limiting Factor* would land in the center of the Western Pool, scoot along in a straight line, then climb up and along one steep wall of the trench. It would be a breathtaking ride.

In previous dives, I've been to places where the ocean floor is being created, but Challenger Deep is the opposite case where tectonic plates are coming together. As a geologist, I can literally get to the bottom of things. —Dawn Wright

Dive prep

No specific training is required to dive in the submersible, but, on the other hand, you need to be in pretty good shape to handle the physical demands of the expedition. The *Pressure Drop* is no luxury liner, and the *Limiting Factor* is designed for function, not comfort. Just getting around the ship and lowering yourself into the sub requires a certain amount of strength and agility, as Dawn explains:

> Well, for starters, the DSSV *Pressure Drop* has five decks connected by steep internal and external stairwells with handrails. There is no elevator, so you must negotiate the stairs (ladders) to move around the ship and be fit enough to climb and descend those stairs multiple times each day. There is also upper-body fitness required to open and close the standard steel waterproof hatches throughout the ship.

> And to get into and out of the *Limiting Factor* requires a bit of flexibility. The submersible is not permitted to launch from the deck of the *Pressure Drop* with occupants already inside (such as with launches of *Alvin*). So, it must be craned into position over the water before we enter the vehicle. We must then step onto the *Limiting Factor* from a short gangway attached to the ship, and we'll be protected with personal flotation devices and a harnessing system, just in case. Once stable and positioned adjacent to the *Limiting Factor* entry hatch, we must remove our harnesses and flotation devices, aided by the ship's crew if needed. Then we must climb down four ladder rungs into the conning tower, reposition, and climb down four more ladder rungs into the *Limiting Factor* personnel sphere, through the main hatch (450 millimeters or 17.7 inches diameter).

In the months leading up to the dive, Dawn (no couch potato to begin with) worked to get into the best physical condition possible. She describes her routine as "upper body and core strength training and flexibility exercises with my physical trainer, along with as much of my beloved road cycling and mountain biking as I can fit in. The cycling always includes significant climbing for cardiovascular fitness and leg strength."

She also had to make other, more personal preparations for being away at sea for 12 days. A major concern was arranging care for her beloved mother, Jeanne Wright, who, at 86, was living with Dawn, in failing health. Once that was set up, there was her golden retriever Riley to consider (no dogs allowed in Challenger Deep!). Luckily, Riley loves being away with her pals at puppy camp, so Dawn knew she'd be fine. Everything was falling into place for Dawn's historic dive.

Risky business

By the end of 2021, the *Limiting Factor* had successfully completed 15 dives into Challenger Deep. With a record like this, you might be forgiven for believing that an 11-kilometer (7-mile) descent into the hadal zone is as simple, and as safe, as hopping on a city bus.

If you believed this, unfortunately, you would be wrong. Submersible dives to such depths, and even lesser depths, are still a risky business, fraught with potential peril. One tiny design flaw or deep-sea mishap could be catastrophic. As Nicole Yamase says, "You're going in a machine that is basically holding your life, and if anything goes wrong, that's it. You would turn to dust immediately because of the 16,000 PSI [pounds per square inch]. The pressure would just crush you immediately."

Not a pleasant image to dwell on, and Dawn certainly doesn't. As a scientist who has been going down in submersibles since 1991, she approaches a dive—any dive—as a professional, doing her job. She even gets a little miffed when people ask her if she ever feels afraid before a dive. Never, she insists—for her, it's just like getting into a car or a plane, and "people don't worry about their safety every time they do that, do they?" (Actually, Dawn, some people do!)

Though it's not something deep-sea explorers like to think about, a catastrophic accident is always a possibility. And, to date, there have been several mishaps and near-misses, including two high-profile incidents: one in 1973 and one in 2023, one with a happy ending and the other, tragically, not.

Pisces III

In August 1973, two British sailors were trapped for three days underwater, at a depth of 500 meters (1,600 feet), in the *Pisces III*, a Canadian commercial sub. When they were finally rescued from the waters of the Atlantic, 241 kilometers (150 miles) off the coast of Ireland, they had only 12 minutes' worth of oxygen left. It was a dramatic (and precarious) international rescue operation that captured the world's attention.

Leading the pack

In addition to her other accomplishments, Dawn has always been an athlete. She started out as a basketball and track athlete, excelling in the long jump, working with her high school coach toward trying out for the 1980 Olympic team. But then she suffered a serious knee injury, which cut short her career as a sprinter. She continued to run, but as a distance runner—not necessarily the best choice for her knees—and earned varsity letters at Wheaton College in basketball, track, and cross-country. By the time Dawn got to UC Santa Barbara (UCSB) for her PhD studies, her knees were pretty much shot (she'd been running marathons on the side!), and she was looking for a new, low-impact sport. UCSB had (and still has) a great road cycling team, so Dawn joined the club, trained like crazy, and was on the squad that took the bronze medal in the 1993 NCCA Collegiate Road Nationals. OK, there were a couple of cycling mishaps along the way—a bad crash and a stolen bike—but both had happy endings, and Dawn still loves to get out on her bike whenever she can.

Dawn on a training ride in Redlands, California.

It began as a routine dive for former Royal Navy submariner Roger Chapman and engineer Roger Mallinson, who climbed into the crew sphere of the *Pisces III* on August 29 to continue laying transatlantic telephone cable on the seabed. "We'd do eight-hour shifts, going along the surface of the seabed at half a mile an hour, setting up pumps and jets which liquefied the mud, laying cable and making sure it was all covered," Chapman told the BBC in 2013. "It was very slow, murky work." Also, Mallison added, the sphere, only 6 feet in diameter, was cramped and uncomfortable: "We had to kneel, with our heads by our knees."

As well as laying cable, the pilots had to manage their life support system, running a lithium hydroxide fan every 40 minutes to soak up carbon dioxide, then letting a small amount of fresh oxygen into the sphere. Luckily, the oxygen tank was full when they set out.

Eight hours later, *Pisces III* was back at the surface as scheduled, waiting for a towline to be attached to lift the sub to the support ship. Suddenly, Chapman said, "We were hurtled backwards and sank rapidly. We were dangling upside down, then heaved up like a big dipper," a roller coaster.

Somehow, the towline had caught on the aft sphere—a small watertight compartment containing the sub's machinery— and wrenched it open. Over a ton of water flooded into the aft sphere, and the sub plummeted like a dive bomber, dials spinning, engines screaming, until it jolted to a halt at 53 meters (175 feet), the length of the towline. It dangled there, terrifyingly, until the line snapped, and the sub plummeted again, taking 30 seconds—the longest 30 seconds of the men's lives—to crash into the seabed at 64 kilometers per hour (40 miles per hour).

Amazingly, the two were unhurt. The sub had landed almost upside down and in a gully, half hidden in the seabed. But they were alive and still in telephone contact with the surface, where a rescue effort was being organized. So Chapman and Mallinson settled in to wait, in utter darkness, freezing cold and wet through, barely talking or moving to conserve oxygen. As the CO_2 levels rose in the sphere, their energy and spirits sank. "But our job was to stay alive," said Mallinson.

Meanwhile, on the surface, ships and subs were being scrambled from as far away as California and the North Sea. But none of the rescue attempts were working: the subs *Pisces II* and *Pisces V* were launched but ran into technical problems of their own, as did the US Navy submersible *CURV III*, designed to pick up bombs from the ocean floor. By midnight August 31, two days after the accident, Chapman recalled, "Our 72 hours of oxygen was up, we were running out of lithium

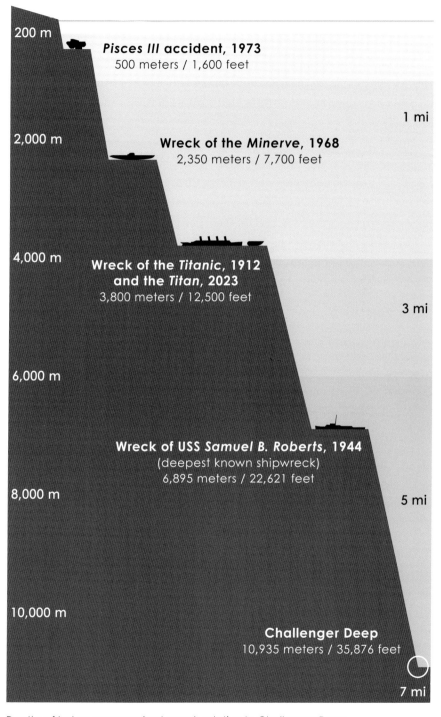

200 m

Pisces III accident, 1973
500 meters / 1,600 feet

1 mi

2,000 m

Wreck of the Minerve, 1968
2,350 meters / 7,700 feet

4,000 m

**Wreck of the Titanic, 1912
and the Titan, 2023**
3,800 meters / 12,500 feet

3 mi

6,000 m

Wreck of USS Samuel B. Roberts, 1944
(deepest known shipwreck)
6,895 meters / 22,621 feet

8,000 m

5 mi

10,000 m

Challenger Deep
10,935 meters / 35,876 feet

7 mi

Depths of lost or compromised vessels relative to Challenger Deep.

hydroxide to scrub the CO_2, it was very manky and cold, and we were almost resigned to thinking it wasn't going to happen."

Mallinson agreed that their situation seemed hopeless. But something that lifted his spirits was the presence of dolphins, which they'd seen on the first day and could hear the whole time through the sub's underwater acoustics, heartening him.

Finally, on September 1, the *Pisces II* and *CURV III* managed to attach lines to the sunken sub, and the arduous process of lifting it began. It was at this point, when they knew the lines were attached, that Chapman and Mallinson shared the only food they had with them—a cheese and chutney sandwich and a can of lemonade. (Probably not the best meal for the rough ride that followed.)

The lifting process was long and brutal, with the men being hurled and rocked about as the sub was hauled up in a series of abrupt stops and starts. When at last *Pisces III* cleared the surface and the rescuers peered in, Chapman said, "Apparently they thought we'd died," because the retrieval had been "so violent."

Divers assist the rescued pilots, Roger Chapman and Roger Mallinson, from *Pisces III*.

By this time, Chapman and Mallinson had been in the stranded sub for 84 hours and 11 minutes. "We had 72 hours of life support when we started the dive," Chapman told the BBC, "so we managed to eke out a further 12.5 hours. When we looked in the cylinder, we had 12 minutes of oxygen left."

Surprisingly, after this near-death experience, both men continued to work in and around submersibles—though Chapman admits he's avoided elevators (confined space, vertical motion) ever since.

Titan

Almost 50 years later, on June 18, 2023, five people happily boarded the Ocean-Gate *Titan* submersible, far out in the North Atlantic, for a dive to view the wreck of the *Titanic*. For three of them—British billionaire and explorer Hamish Harding, British-Pakistani businessman Shahzada Dawood, and his 18-year-old son, Suleman—this would be a once-in-a-lifetime experience, a dream come true. Harding had already descended to Challenger Deep with Victor in March 2021, spending a record 4.25 hours at the bottom, and was now eager to visit the world's most famous shipwreck.

For another passenger—or "mission specialist," as even the fee-paying clients were called—this would be his 38th dive to the wreck. Paul-Henri Nargeolet (affectionately known as "P.H.") was a 77-year-old French scientist, deep-sea explorer, and renowned authority on the *Titanic*. He'd been exploring the wreckage since 1987, soon after it was first located, and had dived there 37 times in five different subs, collecting artifacts for exhibitions. On this trip, he was employed by OceanGate as a guide.

The fifth person onboard that day was Stockton Rush, the cofounder and CEO of OceanGate, a private research and ocean tourism company that had built the *Titan* and offered deep-sea dives to those who could afford the hefty price tag. Rush was a wealthy entrepreneur and adventurer with a background in aerospace engineering. "I had come across this business anomaly I couldn't explain," he told an interviewer. "If three-quarters of the planet is water, how come you can't access it?"

Rush himself had piloted the sub on most of the 13 dives it had completed to the *Titanic*; many more had been aborted for technical reasons. If everything worked perfectly, the *Titan* would take a leisurely two and a half hours to descend to about 3,840 meters (12,600 feet/2.4 miles), spend four hours touring the *Titanic*, and ascend at the same slow pace.

At 7:30 that June morning, the five voyagers crawled into the sub, settling onto the floor of the pressure hull with their backs against its curved walls. (If you thought the airlines were bad, your $250,000 ticket on the *Titan* didn't even buy you a seat. ...) The hatch clanged shut and was bolted from the outside—a departure from industry standards. The *Titan* slipped beneath the waves.

And that was the last time any of those passengers were ever seen alive.

One hour and forty-five minutes into the dive, communication with the sub was lost. Unlike the *Limiting Factor* and other research vessels, which rely on dual-redundant acoustic modems, the *Titan* depended solely on text-based communications with the support ship. The comms were often down, so at first there was no great concern; if the outage lasted more than an hour, the protocol was for Rush to abort the dive and resurface.

Perhaps that's why the crew on the support ship didn't report the sub missing for eight hours, hoping the *Titan* would pop up somewhere unannounced. But once the Coast Guard was alerted, a massive international search-and-rescue operation was launched, racing against the clock: if the sub was still intact (big if), those aboard would run out of oxygen in 94 hours.

A massive media frenzy accompanied the search, with audiences around the world breathlessly following each development, speculating wildly, and ghoulishly counting down the hours. The all-out effort fueled hope that the men might still be found alive. But those in the tight-knit community of deep-sea scientists and engineers guessed immediately what had happened. For years, they had been raising the alarm about the *Titan*'s design flaws.

Rush saw himself as an innovator, a disrupter who didn't want to be bogged down by bureaucratic rules and regulations. For this reason (and to save money), he chose not to have the sub "classed" by a marine-certification agency, as the *Limiting Factor* had been, to meet the highest safety standards. His clients were required to sign waivers acknowledging that the vessel was experimental and unclassed.

Rush's "innovations" included shortcuts and cost-saving measures that dismayed experts in the field, so much so that dozens of them signed a letter to OceanGate in 2018 expressing "unanimous concern" about the sub's design. Among the many red flags: the lack of an emergency radio beacon; the lack of a recovery system that could be launched from the support ship; an off-the-shelf Sony PlayStation 3 as the sub's sole control system; the cylindrical shape of the pressure hull; and, most importantly, Rush's decision to use carbon fiber instead of steel or titanium for the hull.

OceanGate's *Titan* submersible.

The *Limiting Factor*, like most deep-ocean submersibles, has a spherical hull made of titanium, the strongest and most resilient material for this purpose. But there's a trade-off between strength and weight: the weight of the titanium limits the size of the hull and, hence, the number of people it can accommodate. Rush wanted to take as many clients as possible on his *Titanic* tours, and so OceanGate made the risky decision to build the pressure hull out of carbon fiber.

Carbon fiber is strong and significantly lighter than titanium or steel, but there's a reason it's not generally used to build subs: it breaks down slowly under deep-sea pressure. It might be good for 20 dives, or 100, but on the 21st, or the 101st, it could unexpectedly fail. OceanGate was aware of this risk and installed an acoustic monitoring system to "listen" for sounds of stress from the hull. But what if it picked up a warning sound just seconds before the hull failed?

On day 4, the frantic search for the missing sub was called off. Just before noon on June 22, the US Coast Guard announced that a remotely operated vehicle had located debris from the *Titan* on the seafloor, not far from the bow of the *Titanic*. According to the Coast Guard, the debris field, which included the sub's nose cone, was "consistent with the catastrophic loss of the pressure chamber"—in other words, the sub had imploded instantly on the way down. All aboard were believed to be lost.

It was a tragic end to what had begun as a grand adventure. And experts have now concluded that it probably wasn't the carbon fiber itself that caused the hull to fail but the use of three dissimilar materials—carbon fiber, titanium, and Plexiglas—glued together, expanding and contracting at different rates. Whatever the exact cause, lives were needlessly lost in a disaster that knowledgeable people had been trying to prevent for years. Among the explorers Rush had invited to tour the *Titanic* with him was Victor Vescovo, who turned him down. He wanted nothing to do with OceanGate's operation.

Unlimited operations

As we know from chapter 1, Dawn completed her dive safely and successfully, thanks to the world-class engineering and materials of the *Limiting Factor*, and the skilled support crew on the *Pressure Drop*. According to its designers, the "defining feature of the *Limiting Factor* is the 90-millimeter-thick [3.54 inch] pressure hull. Having been machined to within 99.933% of true spherical form, it is testament to the precision engineering required to develop a certified, reusable, full-ocean-depth submersible."

As the first human-occupied vehicle capable of completing "unlimited operations" to the deep, the *Limiting Factor* has, to date, maintained a flawless safety record. No mean feat, when you consider the crushing pressure it has to withstand—comparable, Dawn says, to the weight of 50 fully loaded jumbo jets piled on top of a person.

Immediately before Dawn's dive, in July 2022, Victor's team achieved another first: finding and diving to the world's deepest shipwreck, USS *Samuel B. Roberts*,

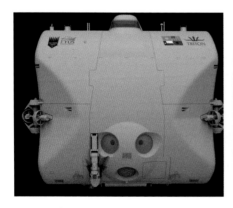 

Design of the *Limiting Factor*.

The wreck of USS *Samuel B. Roberts*, 6,895 meters (22,621 feet) deep.

lost in World War II. The *Sammy B.*, as it's fondly called, went down in October 1944, during the Battle off Samar in the Philippine Sea. Returning to the Philippine Trench to search for it, Victor expressed only about "50 percent hope" that he'd be able to locate the wreck, sunk during fierce fighting between the US Navy and Japan. Of the 224-person crew, 89 were lost that day.

For this search, the *Limiting Factor* deployed the sidescan sonar system that Dawn would later test at full ocean depth. But when the first two search dives for the *Sammy B.* found nothing, Victor's team gave up and began scanning for a smaller aircraft carrier lost in the same area. And then, bingo: the ghostly warship was spotted at a depth of 6,895 meters (22,621 feet), battered, punctured, but still largely intact.

"I was so incredibly happy to find her," Victor said at the time:

> [T]o see the intact hull, her weapons still loaded with live ammunition, her broken mast and shattered stern—it was like seeing history come alive. We were the first people to see her since she sank below the waves that day [October 25] in 1944, and it was such a privilege to be the ones that found her.

He added that the crew of the *Sammy B.* was "extraordinarily brave," and he hoped the discovery of the wreck would provide "an opportunity to retell her heroic history."

Dawn's dive—in pictures

After finding the *Sammy B.*, the *Limiting Factor*'s next mission was Dawn's dive to the Western Pool, scheduled for July 12, 2022. The weather forecast was favorable, and this time, the stars aligned for Dawn Wright to make history.

In chapter 1, we followed the timeline of Dawn's dive, but now let's take a closer, more personal look at her journey, an album of some of Dawn's favorite images from her expedition—an inside look at the life of an explorer at sea.

The schedule called for Dawn to fly to Apra, Guam, on July 9, where she would board the *Pressure Drop* for the trip to Challenger Deep, a distance of 210 nautical miles, about 20–22 hours at sea under favorable conditions. Unfortunately for Dawn—an eminent oceanographer who's spent most of her professional life at sea—she suffers from seasickness. Most people probably would have quit oceanography after the first bout or two, but Dawn says the sickness lasts a couple of days, and, given that no medications have ever worked for her, she just toughs it out.

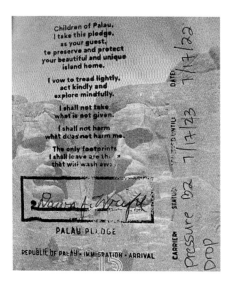

The passport stamp of Palau includes a pledge encouraging visitors to preserve and protect the island.

Jeanne Wright, 1935–2021.

Thankfully, this trip was different. The expedition leader, Rob McCallum, supplied all those in need onboard with what, for Dawn, was a miracle drug—cinnarizine (a calcium channel blocker, manufactured as Stugeron), which has a different mechanism of action from dimenhydrinate (manufactured as Dramamine). This worked for Dawn—not a moment of seasickness, and no side effects, for the entire expedition. Another first!

After her dive on July 12, Dawn would then continue on the *Pressure Drop* to the Yap Trench, to support the dive of Sesario Sewralur, son of the famous Yapese navigator Mau Piailug, a teacher of traditional, noninstrument wayfinding methods for ocean voyaging. Like Yamase's dive, Sewralur's would be deeply meaningful, given his ancestral and familial connections to the ocean in Yap.

Next stop, Palau, where the former president of the Republic of Palau, Tommy Esang Remengesau Jr., widely recognized for his leadership in ocean conservation, would dive into the Palau Trench. And then, on July 20, Dawn would fly back home from Palau.

Before she left, though, Dawn suffered a profound personal loss. Her beloved mother and best friend, Jeanne Wright, passed away on December 7, 2021—Pearl Harbor Day. As Dawn wrote in her online memorial, her mother "led a life of unflinching integrity, continual search for truth, wisdom, and excellence in all things, and unselfish service to God and to others. Hers was truly a purpose-driven life." Perhaps Dawn's own purpose-driven life was Jeanne Wright's most lasting legacy.

Dawn's packing list

Dawn was still grieving when she dived to Challenger Deep, but, she says, she felt her mother's presence watching over her. Also watching over her was a much goofier friend: her Snoopy, which Dawn, as a self-confessed "Snoopy groupie," takes down with her on every dive.

Among the other personally meaningful items that Dawn packed for her dive were some tiny toy sea creatures, belonging to a colleague's three-year-old daughter. The three-year-old, an oceanographer in the making, had asked Dawn to take them down to the deep with her. Dawn also found space in her pack for a book of poems by a friend, Dr. Sian Proctor, the first civilian astronaut to pilot a spaceship and the first Black woman to do so. Proctor's book includes a poem about Challenger Deep, which Dawn read into the camera when they reached the bottom. And, of course, Globie, the Esri mascot, went along for the ride.

Snoopy groupie

Dawn's connection to the *Peanuts* character dates from when she was a child, reading the daily comic strips in the newspaper. She fell in love with Snoopy then, but what really "flipped the switch" for her was a special gift from her mother. For Dawn's eighth birthday, her mother redecorated Dawn's bedroom, going all out with *Peanuts* posters and even a new set of *Peanuts*-themed bedding. As if that weren't enough, Snoopy was also drafted as the mascot for the 1969 Apollo 10 mission, helping to boost public enthusiasm for the space program—Charles M. Schulz even drew Snoopy on the moon. The call signals for Apollo 10's command module and lunar module were "Charlie Brown" and "Snoopy," respectively. Inspired by the high-flying beagle in the 1960s, Dawn has been a proud member of the Peanuts Collectors Club ever since.

Snoopy kept watch
for Dawn over her
Starboard dashboard.

Dawn also took some practical gear, including a flashlight and extra pairs of socks, because the sub isn't heated, and it gets down to 4–7 degrees Celsius (40–45 degrees Fahrenheit) in there. She and Victor both brought along red beanies, which they were planning to wear in honor of Jacques Cousteau and his signature headgear—but somehow that never happened. Victor didn't wear anything on his head for this dive, and Dawn kept her Esri cap on for the whole ride. The most important item, in terms of Dawn's safety, was her blue jumpsuit—a flame-retardant flight suit, certified by NASA, to protect her in the unlikely event of a fire in the sub.

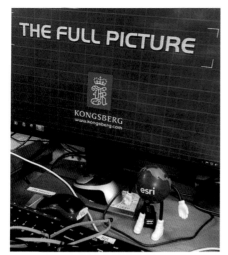

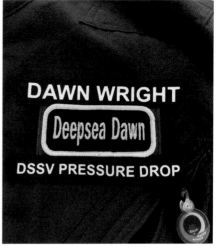

Globie, the Esri mascot, in front of the sonar monitor. Dawn gave Globie to the expedition's mapper, Rochelle Wigley of Map the Gaps and the University of New Hampshire, to show appreciation for her work during the expedition.

Dawn's name patch on her diving suit. The medallion on the zipper was to honor the Black in Marine Science organization.

Dawn (*left*) and Kate Wawatai (*right*) practicing in the sub.

Onboard prep

The day before the dive, Dawn practiced getting into and out of the sub. She wanted to get a feel for where the rungs were located and where to turn and step when lowering herself down into the crew sphere.

During this practice session, Dawn and her teammate Kate Wawatai climbed all the way into the sphere, where Kate gave Dawn an instrument panel and safety briefing. Wawatai, a New Zealander, is the first female pilot of *Limiting Factor* and, to date, the only female technician on the sub team. She hopes to one day pilot the sub to the deepest part of the Kermadec Trench, a sacred place in Maori history and culture.

Touchdown!

Dawn and Victor touch down in Challenger Deep.

After reaching the seafloor at an initial depth of 10,904 meters (35,774 feet), an exuberant Dawn and Victor took time for a touchdown selfie.

During their dive, Dawn and Victor explored over 1,500 meters (4,921 feet) of the southern wall in the Western Pool of Challenger Deep, an area that had never been surveyed before. Apart from the infamous beer bottle and the outcropping of chunky basaltic rocks that Dawn nicknamed Flintstones' Quarry, what else did they see?

Hydroids growing out of basaltic rocks in Challenger Deep.

Anemones growing out of a basaltic rock formation in Challenger Deep.

Dawn's dive dedication

On July 12, 2022, when Dawn became the first Black person to descend to Challenger Deep, she read these words on the sub, dedicating her dive to the following:

- Sian Proctor, who in September 2021 became the first Black woman to pilot a spacecraft (*Crew Dragon*), as a civilian astronaut on the SpaceX Inspiration 4 mission. She's also a fantastic poet, artist, and community college geology professor. Her poem "Seeker":

  Why Go?

  Because I can

  It is what humans do

  We explore, We observe, We learn

- Full Circle Everest, the first all-Black expedition to summit Mount Everest, in May of this year [2022].
- Evan B. Forde, the first African American scientist to conduct research in a submersible, the *Nekton Gamma*, and then in *Alvin*, and then in *Johnson Sea Link*.
- Ayesha McGowan, the first female African American professional road cyclist, currently on the Liv Racing Xstra team.
- Albert José "Doc" Jones, founder in 1959 of Underwater Adventure Seekers, the first club for African American scuba diving enthusiasts and, to this day, the leading light of Black scuba diving in US waters.
- My NEW friends, the Friends of the Mariana Trench, and their efforts to establish a national marine sanctuary in the Northern Mariana Islands, including this trench.
- And finally, to my beloved mom, Jeanne, who was fitted with her angel wings in December of last year [2021], but who, despite her motherly fears, always encouraged and celebrated each and every one of my expeditions at sea.

Dawn says, "The strangest life-forms that we encountered during the dive were these hydroids growing out of the basaltic rocks. These hydroids were identified by Dr. Alan Jamieson of the University of Western Australia." Deep-sea hydroids are creatures that attach themselves to rocks and, as this image shows, they look like white worms coming out of the rock.

At such extreme depths, not many creatures can survive. Among them, though, are these white, tube-dwelling anemones of the genus *Galatheanthemum*—life-forms of otherworldly beauty, and the only genus of this organism to be found so deep.

Back on top

After her successful dive, Dawn received a call aboard the *Pressure Drop* from the White House. On the line from the White House Office of Science and Technology Policy, via satellite speaker phone, were two eminent scientists: Dr. Alondra Nelson, deputy director for science and society, and Dr. Jane Lubchenco, deputy director for climate and environment. They congratulated Dawn, wanted to hear all about her dive, and asked her to discuss the scientific significance of Challenger Deep.

Dawn aboard the DSSV *Pressure Drop*, on a call with Dr. Alondra Nelson and Dr. Jane Lubchenco from the White House Office of Science and Technology Policy.

Of course, Dawn, being Dawn, also had some fun aboard the *Pressure Drop* as the expedition continued to the Yap Trench and the Palau Trench. "Is it fun building with LEGO Group bricks at sea?" she wrote. "What a question?! OF COURSE, IT IS!"

Bathymetry maps and LEGO bricks at sea.

Why LEGO?

Surprisingly, for someone who's an official "AFOL" (adult fan of LEGO), has been featured as a LEGO minifigure, and maintains a spreadsheet listing her impressive number of LEGO builds, Dawn didn't really play with LEGO as a child. In fact, her "deep, dark secret" is that she only got turned on to LEGO in a big way when she was preparing for her qualifying exams as a doctoral student. During that stressful time, sequestered and studying hard, she found building with LEGO a good way to relax. Also, Dawn says, there's a "symmetry and success" to LEGO, because you know that, if you follow the instructions, you're going to get a good result. Of course, her very first build was a pirate ship!

These days, Dawn likes to describe LEGOs as "a wonderful analogy to open science." As Dawn says, "You can continually build on top of what others have accomplished. Through publishing and open sharing, you give your bricks to others and interoperate with bricks for reuse, especially with others who have not had the chance. That's what I want the world to be like: people coming together and helping each other despite their differences."

Two of Dawn's favorite LEGO builds are shown here.

Dawn's favorite build of all time: The glorious Imperial Flagship (Kit #10210, 1664 pieces). 1977 Heezen and Tharp *World Ocean Floor Panorama* in the background.

LEGO Research Vessel Build. Port side: female captain, Deepsea Dawn scientist, and female deep diver minifigures.

As the expedition came to an end, Dawn posed with two key teammates for a final photo, marking their joint achievement. "This is one of my favorite photos," Dawn says. "Super submarine tech Kate Wawatai, explorer and owner extraordinaire Victor Vescovo, and I celebrate at the conclusion of the expedition!"

By the time Dawn returned to dry land, news of her dive had spread around the world. A TV crew had been on the *Pressure Drop* to document her journey, and, once that segment aired on CBS, interview requests kept pouring in. Dawn was featured in all kinds of media, from podcasts to scientific journals to *Essence* magazine. But in all the interviews about her dive, one question kept coming up: *Why? Why* do we explore the deep? *Why* do we map the ocean floor?

As you might imagine, Dawn and other ocean scientists have a pretty good answer to that question. We do it because, as humans, our lives depend on it.

Kate (*left*), Victor (*center*), and Dawn (*right*) at the end of the successful expedition.

Chapter 6

WHY WE MAP THE DEEP (AND WHY YOU SHOULD CARE)

" Mapping is really the science of our world. —Dawn Wright

"D o you like to breathe?" This is what distinguished oceanographer and Dawn's longtime mentor Sylvia Earle ("Her Deepness") sometimes asks skeptics who wonder why we study the deep. After all, they say, there's nothing down there—just some sand and rocks, right? Let's go to Mars instead!

No, Earle patiently explains: If the ocean and its organisms weren't in good health, we humans wouldn't be either. For starters, about 50 percent of the oxygen we breathe comes from organisms in the ocean. Marine photosynthesizers, such as seaweed and phytoplankton, use carbon dioxide, water, and energy from the sun to make food for themselves. In the process, they release oxygen into the atmosphere. If they weren't doing this, we wouldn't be breathing.

This cycle takes time. Most of the oxygen produced by the ocean is directly consumed by the microbes and creatures that live there, or consumed when organic matter falls to the seafloor. So about half the oxygen we breathe now comes from the slow accumulation of oxygen in the atmosphere over millions of years.

As well as providing us and other species with essential oxygen, the ocean plays a critical role in regulating the planet's climate. It also buffers our excess carbon (so far, anyway), absorbs surplus heat (so far), produces a wealth of resources, feeds millions of people, and keeps the global economy afloat. At the same time, it can unleash destructive forces—as we've seen, most earthquakes, volcanoes, and tsunamis originate in the deep.

Ocean waters cover 70 percent of our planet, but if we don't fully understand how the ocean works, we can't protect it (or ourselves). And if we can't protect it, all species on Earth, including our own, are at risk. Sylvia Earle has sounded the alarm: "We need to act now," she warns, "because if there's no blue, there's no green, which means there will eventually be no humans. No kidding!"

That's why scientists like Dawn Wright, Cassie Bongiovanni, Kathy Sullivan, Nicole Yamase, and Sylvia Earle have dedicated their lives to exploring the ocean, studying its biology and geology, and mapping its depths.

Instead of the flat blue surface of earlier maps, or imaginary monsters of the deep, today's maps show us the hidden face of our planet: a topography scarred and riven by its formation; a dramatic terrain of volcanoes, trenches, and fracture zones; a hotbed of geologic activity. This geologic activity, Dawn believes, is "the most exciting, the most perplexing, and the most important" to study.

The most important? Yes, Dawn says, and here's why.

Transformation and destruction

For starters, Dawn says, what happens in the ocean doesn't stay in the ocean. We now know that the planet's surface is made up of tectonic plates, major and minor, that ride on top of Earth's mantle—essentially a layer of hot roiling rock. The plates fit together like a jigsaw puzzle, but, because the mantle is moving, they don't stay put: they're constantly shifting, sliding, and grinding into one another. The seams between them, where they collide or push apart, are known as faults. Just like on land, where a sudden slip along a fault can cause an earthquake, undersea faults are hot spots of geologic activity, generating earthquakes and volcanic eruptions on the Earth above. But high-resolution seabed maps can help scientists predict catastrophic events like these before they happen.

Also, Dawn explains, "Tsunamis come about as a result of a movement on the floor of the ocean that's expressed in a fault. When you have that disruption on the ocean floor, the water above the ocean floor gets severely disrupted as well, and generates these large, long-wavelength waves that are essentially the

A Spilhaus projected coordinate system map showing the major tectonic plates.

tsunamis." That's why we need to monitor those fault zones, but even more importantly, we need to understand the shape of the seafloor in coastal areas where tsunamis occur. Though a tsunami is a terrifying mass of water, how it behaves and how fast it travels is influenced by what's below: the shape of the coastline and topography of the ocean floor.

According to NOAA, tsunamis "have killed hundreds of thousands of people worldwide and caused billions of dollars in damage. They are equal opportunity destroyers: no coastal area in the world is entirely safe from them." But if we have better maps and, hence, better models, we can more accurately predict the impact of tsunamis and—ideally—save lives through more effective warnings (when, how, and where to evacuate).

Energy

"Another very, very important reason" to map the ocean floor, according to Dawn, is that "we're looking more to the ocean now for energy—wave energy, for instance, or wind energy. And we are siting wind turbines on the ocean floor. So we need to know what the ocean floor looks like—which areas are flat, for instance, and what is the composition of those areas. Is it sandy? Is it silty? Is it rocky? What are the best places for us to anchor that type of alternative energy infrastructure?"

Siting wind turbines involves a host of other considerations, notes Dr. Richard Spinrad, head of NOAA in the Biden administration. Among those is ensuring that the turbines won't harm endangered marine species, such as the North Atlantic right whale. With the right geospatial data, Spinrad says, we can begin to answer crucial questions, such as "How do we ensure that while building out this renewable energy source we are also preserving the right whale? How are we going to know that the wind will be there in 30 years?" Accurate maps and spatial models can guide scientists to make the best decisions for our energy needs as well as for the ocean ecosystem.

Also, as offshore renewable energy becomes increasingly important in the race to replace fossil fuels, new questions arise about access and rights to areas of the ocean. For both economic activity and conservation efforts to succeed, we need well-defined ocean territories. Who will reap the benefits from that part of the ocean? Who will take care of it? Who will protect it from overexploitation? The first step toward responsibly managing any area of the ocean, scientists say, is defining it—i.e., mapping it.

Communication connections

Do you spend too much time on social media? Lose yourself down research rabbit holes? You probably don't ever think about the ocean floor while going online, but Dawn suggests you spare a thought for the infrastructure down there. More than 1.3 million kilometers (about 800,000 miles) of underwater data and communications cables traverse the seabed, connecting distant continents (and billions of humans) in real time.

"Our maps show that 99 percent of all of our internet traffic is carried by these submarine cables," Dawn says. This means that pretty much everything that streams into our lives today—"from podcasts to Pokémon Go accounts to a WhatsApp group to all of the streaming videos"—depends on this undersea

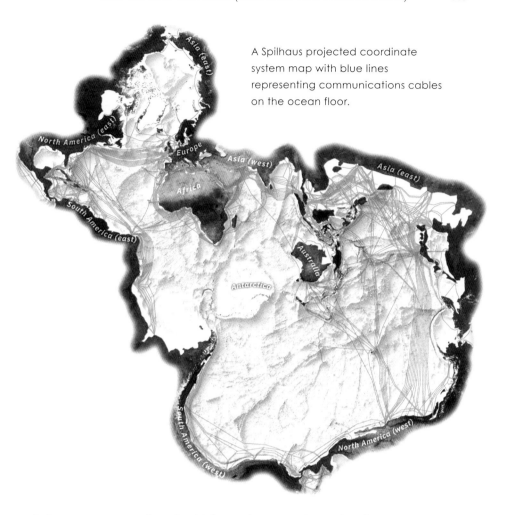

A Spilhaus projected coordinate system map with blue lines representing communications cables on the ocean floor.

infrastructure. As a fan of British murder mysteries on her favorite streaming service, Dawn says she's "thankful to the ocean floor for that!"

Although submarine cables are strongly reinforced, they are still liable to failure—especially if they're placed on areas of the seabed that are likely to shift (or blow), like fault lines and volcanoes. With high-quality seafloor maps, rather than best guesses, we can avoid laying these all-important cables in hazardous zones.

Habitat

Many of us, wherever we live, enjoy fish and other types of protein from the sea—and communities around the world depend on vital marine environments for their survival. Some 60 million people are employed in fisheries worldwide, while fish and other seafood products provide essential nutrients for more than

3 billion people on the planet. Much of that much needed protein is swimming around the ocean floor, especially near seamounts, so Dawn says we need accurate mapping "to monitor and protect the habitat of those fisheries."

If scientists have a clearer picture of life beneath the surface, they can do a better job of managing marine fish resources, making sure that local fish stocks aren't depleted, and protecting vulnerable species from overexploitation. Almost 90 percent of global marine fish stocks are already fully exploited or overfished, so this is an urgent problem that accurate ocean mapping can help address.

Of course, protecting marine habitats has value over and above its immediate utility to humans. As anyone who's ever watched a nature documentary knows, there's a stunning world of biodiversity down there, from the sunlight zone to the hadal depths, from microscopic plankton to the gigantic blue whale. An estimated 50–80 percent of all life on Earth is swimming, crawling, and wriggling beneath the surface, and scientists have identified only about 250,000 species so far—with as many as 2 million more still to go.

Those are mind-boggling statistics: so many creatures yet to be discovered and identified, so much work for oceanographers still to do, so many marine habitats to be protected—because, as we know, the health of the planet depends on the health of the ocean.

The recently identified *Psychropotes longicauda*, nicknamed the "gummy squirrel," found at depths of 5,100 meters (16,732 feet) in the Pacific Ocean. A type of sea cucumber, it's bright yellow, 80 centimeters (32 inches) long, with a red underbelly and 18 red "lips" (feeding palps). What other weird and wonderful creatures are down there?

Resilience

Coastal inundation (flooding) poses a significant threat to millions of people around the world, especially in small island nations, and that threat will keep rising as sea levels rise. The global mean sea level has risen about 21–24 centimeters (8–9 inches) since 1880, a trend that is accelerating as the Earth heats up, melting huge glaciers and ice sheets. This is bad news for people living in densely populated coastal areas; in the US alone, that's almost 30 percent of the population.

If scientists have good information about the shape of the seafloor and the nature of the coastline, they can more accurately predict how inundation will affect low-lying areas—and take measures to protect those coastal communities before it's too late.

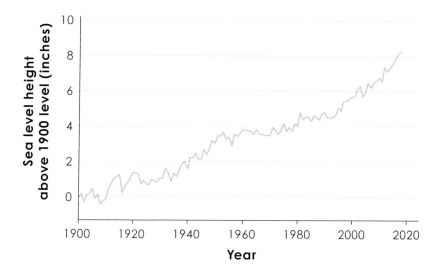

Coastal tide gauge data graph that shows sea level change.

Safety

" When a ship runs aground, it's a very bad day for the environment, for the economy, and, of course, for the captain, too. —DR. LARRY MAYER, director of the Center for Coastal and Ocean Mapping at the University of New Hampshire

At any given moment, tens of thousands of ships are at sea, many of them transporting consumer goods. If even one of them runs into an undersea obstacle,

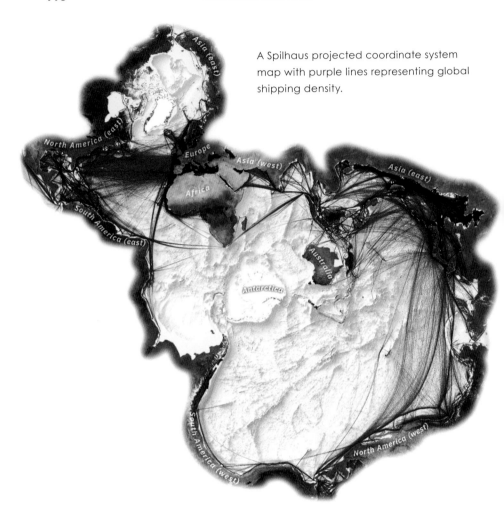

A Spilhaus projected coordinate system map with purple lines representing global shipping density.

spewing fuel or other chemicals, it can do serious harm to the environment. If that vessel is a bulk carrier or oil tanker, the environmental effects could be catastrophic (as we saw, for example, in the 1989 *Exxon Valdez* disaster).

As ocean traffic continues to increase, seabed maps will play an increasingly important role in high-seas safety, giving us better insight into what's under the surface and, hence, better tools for navigation. Accurate maps will help safeguard all those vessels crisscrossing the ocean, the global supply chain that depends on them, the crews that operate them, *and* the sensitive environments through which they sail.

Also, if a ship or a plane goes down in the ocean, we need adequate maps to locate it. "If we have additional accidents or emergencies at sea, how are we going to find things?" Dawn asks. "We still have not found MH370, that airplane that was lost. There was never going to be a very quick way to find or to rescue the OceanGate *Titan* submersible. And those are just two individual examples"—two examples of how, without high-resolution ocean mapping, we're searching randomly in the deep.

The mystery of Flight MH370

It's one of the greatest unsolved mysteries of our time: How can a Boeing 777, the world's largest twin-jet airliner, simply disappear from the skies, with 239 souls on board, and never be found?

On March 8, 2014, Malaysia Airlines Flight 370 (MH370) disappeared from air traffic control radar during a flight from Kuala Lumpur, Malaysia, to Beijing, China. After the aircraft transmitted its final automated position report and made an unscheduled change in direction, radar tracked the aircraft across the Malaysian peninsula, until it was out of range. Then, analysis of satellite data revealed that MH370 had flown for about another six hours over the southern Indian Ocean. After that—nothing. No trace.

To date, even after a yearslong search—the most expensive and high-tech search in aviation history—the wreckage of MH370 hasn't been found. Theories, and conspiracy theories, abound, but they do nothing to assuage the pain and unanswered questions of the passengers' loved ones.

Why has the plane been so difficult to find? One reason may be geology—the rugged topography of the Indian Ocean floor. The best available data suggests that the aircraft entered the water close to a long, narrow arc in the southern Indian Ocean known as the 7th Arc (an arc of possible aircraft positions, equidistant from the Indian Ocean Region satellite, where the aircraft made its final transmissions). It's a remote part of the southern Indian Ocean, about 2,000 kilometers (1,242 miles) from Perth, Australia, a six-day journey by ship. That area of the seabed was largely unexplored at the time the search began, making high-resolution mapping an essential first step.

At the request of the Malaysian government, Australia led the search operation, starting with a two-month aerial search of the surface, with no results. Then the Australian Transport Safety Bureau, working with the governments of Malaysia and China, undertook an intensive underwater search in the southeastern Indian Ocean.

The search unfolded in two phases: Phase One, a bathymetric survey, provided a detailed map of the seafloor topography in the search area. This, in turn, was used to guide Phase Two, the underwater search. The underwater search used sidescan and multibeam sonar equipment mounted on towed and autonomous underwater vehicles (AUVs) to produce high-resolution maps of the seafloor. As the vessels moved across the search area—gathering data 24/7—experts reviewed the sonar data in real time to flag any anomalies that could be an aircraft debris field, warranting closer examination by an AUV. Surveying an area of about 279,000 square kilometers (107,722 square miles) between June 2014 and June 2016, this mission produced the most detailed underwater map ever created.

And yet, sadly, it failed to find the plane. The underwater search was suspended in January 2017. This massive mapping effort located four shipwrecks and an array of previously unknown geologic features—vast seamounts, deep trenches, towering ridges, volcanoes—but no trace of MH370. In all, 447,397 square kilometers (172,741 square miles) of bathymetric data was collected in the search area and another 695,235 square kilometers (268,432 square miles) of data during transit to and from the search area. This data, at least, represents a benefit from an otherwise unfathomable tragedy, maps that will inform scientific explorations of the southern Indian Ocean for years to come.

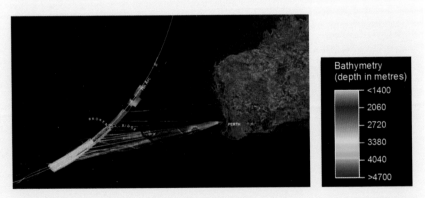

The search area in the remote southern Indian Ocean is now among the most thoroughly mapped deep-sea regions on the planet.

To date, more than 20 items of debris have been analyzed by the current Malaysian investigation team. Items thought to be from MH370 have been found along the east and south coast of Africa, the east coast of Madagascar, and the islands of Mauritius, Réunion, and Rodrigues in the Indian Ocean. And the locations of confirmed debris are consistent with drift modeling indicating that MH370 did, in fact, go down in the search area.

That slice of the Indian Ocean has now been extensively mapped, contributing to our global store of knowledge about the seabed. But the aircraft itself has yet to be found, its location still a mystery for science to solve.

Climate emergency

In 2023, an interviewer asked Dawn: "What if the seafloor isn't totally mapped 30, 40, 50 years from now? What are the consequences of not mapping that seafloor?"

According to Dawn, the consequences could be dire. "We are going to be putting our public safety at risk," she said, "by not having the proper understanding of not only tsunami run-up but storm run-up along our coasts. We are going to be running the consequences of not fully understanding sea level rise and shoreline change. We are going to have a hard time with the security of our ports. We could be losing valuable fisheries forever." And, she added, "The longer that it takes us to get to close to 100 percent [of mapping the seabed], the longer we are going to be playing with fire, so to speak."

Those possibilities are alarming enough. But in an article titled "To Save Earth's Climate, Map the Oceans," Dawn summarizes what, for her, is the most compelling argument of all:

> Climate change is the most basic and urgent reason to map the ocean as quickly as possible. Healthy oceans play an outsize role in minimizing climate change because they capture carbon emissions. But this capacity has limits. Excess carbon acidifies ocean waters, making life difficult for coral reefs and shellfish (oysters, mussels, snails, and clams). It also lowers the oxygen content of the water, impairing the ability of all sea life to breathe. Human practices that disturb the ocean floor—chiefly trawl fishing—make matters worse by releasing carbon from the ocean floor. Deep-sea mining, if it is allowed to go forward unmanaged, would have a similar effect and further disturb undersea ecosystems.

To measure the progress of climate change and study the ocean processes and human activities that affect that progress, it is essential to assemble a detailed picture of the undersea world. Too many people are still thinking of the ocean as "out of sight, out of mind" and not relevant if they don't live near it. This is a luxury we can no longer afford.

It took almost 100 years to survey all the land on Earth. Scientists are already racing to study the seabed before it's changed forever by pollution, overfishing, and climate change. As Dawn's plea makes clear, we don't have the luxury of another 100 years to map the ocean, bit by bit.

During her dive to Challenger Deep, the DSSV *Pressure Drop* collected nearly four terabytes of data, which scientists at Map the Gaps, Caladan Oceanic, and Esri have processed to add to our global store of information about the deep. But about 75 percent of the seafloor still needs to be mapped to modern standards. Can this really be achieved by 2030, which is the ambitious goal of the Nippon Foundation-GEBCO Seabed 2030 Project? And, if so, how?

Chapter 7

CHARTING A HIDDEN WORLD

" *So many of us have rightfully gazed up at the mysteries far above our heads,
searching the solar system for more life and reasons for how we came to be,
cheering the technological advances that go up, up, up. We need to also glance
down and do more wondering at the water below.* —DAWN WRIGHT

Any oceanographer will tell you that it's much easier to map Mars—225 million kilometers (140 million miles) away—than it is to map our own ocean floor, right here on planet Earth. We've mapped the entire moon, no problem. Mars: 100 percent. But most of our underwater Earth has yet to be mapped—fully mapped, in high resolution. The reason for this—apart from our human tendency to look upward, toward the stars—is that we use different technologies to map Mars and to map the deep.

To map other planets, Dawn explains, scientists use electromagnetic energy, such as light, which travels easily through Earth's atmosphere and space, but not so well through water. Light doesn't penetrate the ocean beyond the first few hundred feet—as we saw on Dawn's dive, as the sub descended through aqua blue, then gray, then black. Because Earth is a water planet, and because acoustic energy travels well through water—think of whale song—this is the best type of energy to use to map the ocean floor. But gathering data from acoustic instruments takes much longer than gathering data from satellites or other types of airborne or spaceborne technology. It's also much more labor intensive. Compared with mapping 70 percent of our own planet, mapping Mars is a piece of cake.

Today, scientists have an array of technologies at their disposal for ocean mapping—multibeam sonar, sidescan sonar, satellites, GIS, undersea robots, and sea drones, to name a few. It's still a long, slow process. But, before that, how did we even begin to build a picture of our underwater world, with all its dramatic geologic features—its seamounts and trenches, its vents and volcanoes—hidden from view? How was it even possible to chart the deep?

It all began with the work of an extraordinary, unstoppable woman, one whose story Dawn often mentions as an inspiration and "guiding light" in her own career. Working for most of her life in obscurity, and without the technology available to mappers today, Marie Tharp was the first scientist to produce a detailed map of the global seafloor—and in doing so, completely changed our understanding of how the Earth was formed.

The map that changed the world

" The history of science is full of all these examples of people who just wouldn't give up. —DAWN WRIGHT

In 1948, Marie Tharp had three degrees—a bachelor's in English and music, a master's in geology, a master's in mathematics—and no job.

As a woman born in 1920, her career options were basically limited to teacher, nurse, or secretary, none of which roles interested her in the least. Her father had worked as a soil surveyor for the US Bureau of Soils, and, as a child, Tharp had accompanied him in his fieldwork, learning about science and mapmaking, learning to pay attention.

For Tharp, as for many women, World War II offered an unexpected opportunity. Soon after the war began, the geology department at the University of Michigan opened its doors to women, since most college-age men were off fighting. Tharp leapt at the chance to earn an accelerated master's degree in geology, which came with a guaranteed position in the petroleum industry.

She worked for an oil company for three years but was looking for work when she landed an interview with the geophysicist W. Maurice "Doc" Ewing at Columbia University. According to Tharp, Ewing had a hard time wrapping his head around her array of degrees but eventually just blurted out, "Can you draft?"

Ewing, who had pioneered the use of sonar with the US Navy, wanted to continue sounding and studying the seafloor; by a stroke of luck, he was about to become director of the brand-new Lamont Geological Observatory in Palisades, New York. He needed people like Tharp, skilled in mathematics, who could also draft and translate seafloor soundings into maps.

Ewing hired her as a research assistant (and the first woman in a scientific role at Lamont). Even though she had to work as a subordinate to her less-qualified male colleagues, Tharp recognized the significance of the project and her role in it: "It was a once-in-a-lifetime—a once-in-the-history-of-the-world—opportunity for anyone, but especially for a woman in the 1940s," she later wrote.

Tharp soon began working exclusively with Bruce Heezen, then a graduate student but already the chief scientist of the lab and leader of the ocean expeditions to collect the soundings. It was a collaboration that would last until Heezen's death in 1977 and would revolutionize our understanding of the Earth's geology, past and present.

At the time, seafloor soundings were made using single-beam echo sounding, which is fairly basic compared with the multibeam sonar systems of today but still a huge advance on the way HMS *Challenger* made its depth soundings—by throwing a weighted rope over the side of the ship, hoping for the best, waiting for it to hit bottom, then laboriously hauling it back up again. With echo sounders, the data points were recorded by a stylus passing across long (*very long*) rolls of paper. The lines on the paper rose and fell to reflect the time it took for the sounders' pings to travel from the seafloor back to the moving ship.

Unfortunately for Tharp, women were not allowed on ships in the 1940s—a hangover, perhaps, from the ancient superstition that women brought bad luck at sea—so she couldn't collect her own data. (She wasn't permitted to join an ocean data-collecting expedition until 1965.) As a workaround, Heezen provided Tharp with reams of depth soundings collected from his own research trips and numerous other sources. Tharp's challenge was to figure out how to represent this data in a meaningful way.

First, she marked the ship's course on a map showing latitude and longitude. Then, by hand, she painstakingly plotted the data points (the depth soundings) onto that map as seafloor profiles, exaggerating their height for visibility. Finally, using her drafting skills and her talent for data visualization, she stylized those profiles into readable maps, applying her own understanding of geology to fill in the gaps between data points. What was revealed was a vast underwater landscape—with its ridges, seamounts, and trenches—as dramatic and varied as any visible topography on Earth.

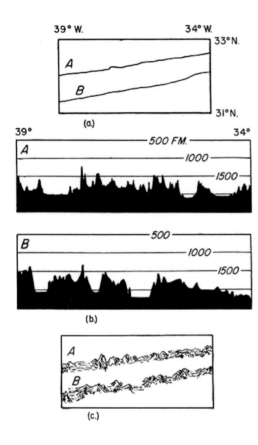

An illustration of Marie Tharp's mapping process: (a) shows the position of two ship tracks (A, B) moving across the surface; (b) plots depth recordings as profiles, exaggerating their height to make features easier to visualize; and (c) sketches features shown on the profiles.

As Tharp toiled away on her map, with new soundings coming in all the time, something odd caught her eye. The profiles showed a gigantic ridge system rising from the seafloor, and in that ridge was a deep V-shaped cleft, lining up along the axis of the ridge. Tharp suggested that the cleft might be a rift valley, like those on land—and was laughed out of the room. Rift valleys are formed when the Earth's crust is pulled apart, and Heezen didn't want to hear it. He dismissed her hypothesis as "girl talk" and told her to go off and recalculate. She did, with the same result.

The reason Heezen was so resistant to Tharp's findings was that, if true, they would confirm the theory of continental drift, which at the time was scientific heresy. This theory had first been proposed by the German scientist Alfred Wegener, who noticed that if you look at a world map, the continents seem to fit together like giant puzzle pieces, with the oceans in between. Perhaps, Wegener theorized, the continents and the oceans were not fixed in place, as had long been believed, but were moving (very slowly) across the surface of the Earth.

Wegener's theory, published in the early 20th century, was widely ridiculed by the scientific community—*Landmasses moving around? What next!*—but Heezen eventually had to accept Tharp's findings. Another researcher at Lamont happened to be plotting the locations of undersea earthquake epicenters on a map that was the same size and scale as Tharp's; when the two maps were superimposed, the epicenters lined up right inside Tharp's rift valley.

Why was this significant? It showed that the seafloor was geologically active, and that, yes, it was indeed moving, spreading apart. Not only had Tharp, through her meticulous work—and the courage of her convictions—discovered Earth's largest physical feature, the 40,000 km (25,000-mile) Mid-Atlantic Ridge, but by identifying the rift valley, she'd rocked the foundations of geology.

Her discovery led to a paradigm shift in the earth sciences and, ultimately, to the development of plate tectonic theory: our current understanding that the Earth's crust is broken up into large rocky plates that move as distinct masses, colliding, spreading apart, causing all kinds of seismic and volcanic havoc. Challenger Deep is one example, formed by two plates crashing into each other, with the heavier one plunging beneath the lighter.

Marie Tharp at her drafting table in Lamont Hall, circa 1961.

After such an earth-shattering discovery (so to speak), you'd expect Marie Tharp to be showered with accolades and honors. No such luck for a woman working in science in the 1940s and 1950s. In fact, Heezen and Ewing published the work under their own names in 1956 and took credit for it. It wasn't until 1959 that Tharp's name appeared on a scientific paper alongside theirs.

When the *New York Times* reported these findings in 1957, the article described the Earth as being "pulled apart." This prompted a spate of letters to Lamont from panicked members of the public, fearing for their safety when the planet split in two. Still, many in the scientific community had a hard time accepting the existence of the rift valley. Jacques Cousteau, for one, was not convinced. So, to prove Tharp wrong, he towed a video camera behind his ship as it sailed over the mid-ocean ridge. Oops: The footage revealed exactly what Tharp had described, an immense underwater mountain range. Cousteau became a convert.

“ *I was so busy making maps, I let them argue.* —MARIE THARP

"We [Tharp and Heezen] were upsetting all the theories," Tharp told an interviewer in 1999. "We learned about 10 things in college, and we had disproved them all."

Although Tharp lost her academic patronage and institutional support when Heezen died, she continued the project, cobbling together support from other sources, working out of her own home, never giving up, producing spectacular map after spectacular map. Her most famous (and beautiful) map, the *World Ocean Floor Panorama*, painted by the Austrian painter Heinrich Berann, was published in 1977. It was based on 25 years of Tharp's work (with Heezen), a true labor of love. It now adorns the walls of many of the world's great institutions of oceanography.

Today, what's remarkable about Tharp's maps is how consistent they are with modern maps of the same areas, which are based on massively more data—satellite and sonar—than was available to Tharp, who used her intellect, skill, and creativity to interpret the data she did have.

One modern legacy of Tharp's panoramas, Dawn says, is Esri's Ocean Basemap, with its carefully chosen colors, saturation, and shading, legible labels, and clear hierarchy of information, inspired by Tharp's meticulous design. Now, with GIS technology, scientists can instantly add their own high-resolution bathymetric data, contributing to our collective understanding of the deep.

The *World Ocean Floor Panorama* by Marie Tharp and Bruce Heezen, painted by Heinrich Berann, 1977.

Dawn herself has been inspired by Tharp's work ever since she first encountered the *World Ocean Floor Panorama* as a grad student at Texas A&M. And, she says, she "fell more deeply in love" with the map once she learned the backstory of its creator. "It helped inspire me to specialize in the study of the shape of the ocean floor and the geological processes at play," Dawn wrote in 2019. "And it stayed with me as my career took off." On her expeditions to the East Pacific as a marine tech, Dawn kept Tharp's 1967 Pacific Ocean panorama at hand, to help her understand the features the research vessel would be drilling into. Even with the detailed bathymetric data available to oceanographers today, Tharp's elegant and eye-opening work remains an essential (and beloved) resource.

Like many women in science, Tharp didn't receive the recognition she deserved until late in her life. That recognition eventually arrived in the form of honors and awards—there is now even a crater on the moon named for Marie Tharp—but, until her death in 2006, what mattered most to her was the work itself. "Establishing the rift valley and the mid-ocean ridge that went all the way around the world for 40,000 miles—that was something important," Tharp reflected. "You could only do that once. You can't find anything bigger than that, at least on this planet."

Ocean mapping today

> ❝ *For the first time, our knowledge of the ocean can approach our knowledge of the land. We can turn the unknown deep into the known deep.*
> —DAWN WRIGHT

Mapping technology has come a long way since Marie Tharp's day, but scientists still rely on acoustic energy to map the deep. Today, acoustic instruments, such as sonar, are deployed in two ways: either at the surface of the ocean, in large arrays from the bottoms of ships and surface drones, or closer to the ocean floor, in portable arrays from deep-diving drones and submersibles. In Dawn's expedition, for instance, multibeam sonar was deployed from the *Pressure Drop* and sidescan sonar from the submersible, collecting both bathymetry (from the ship) and backscatter data (from the sub).

Instruments deployed at the surface can already provide good maps of the ocean floor. As Dawn's good friend and colleague Dr. Vicki Ferrini of the Lamont-Doherty Earth Observatory notes, "Modern acoustic methods can generate hundreds of densely spaced depth measurements in a matter of seconds, and hundreds of thousands of measurements in an hour." This is a huge advance on the hand-plotted maps of Marie Tharp, based on data points from single-beam echo sounders. But, Dawn says, certain places, such as Challenger Deep, need to be surveyed in much greater detail to guide the landing of submersibles, deep-diving robots, and other instruments. The better our data, the better our understanding of the ocean—and the more effective our efforts to protect it.

The good news, according to Dawn, is that now we have the right technology to study—and save—our ocean. "Robotics and sensors and other instruments are creating tons and tons and tons of beautiful data," she reports, and when processed by modern mapping software, this mass of data yields 3D images of the world below. What we've discovered so far is stunning.

Tens of thousands of undersea volcanoes. A coral reef taller than the Empire State Building. An undersea mountain as tall as three Eiffel Towers. The otherworldly white towers of the "Lost City" hydrothermal vent in the Atlantic. The site, in the Pacific, of the first underwater volcanic eruption ever witnessed by humans. And, on the downside, DDT contamination across an area of the seabed as large as San Francisco.

Hugely important information about our planet, hidden from view. What else is down there that we don't know about?

Tubeworm Barbecue

During the 1991 expedition when Dawn made her first *Alvin* dive, her fellow scientists happened upon the aftermath of a recent deep-sea volcanic eruption. Expecting to see an abundance of life around the hydrothermal vents in their study area, they found instead a seafloor covered in (very) recent lava flows. All the creatures around the vents had been wiped out, leaving behind a strange scene of carnage, which came to be dubbed the "Tubeworm Barbecue."

The Barbecue site, known as 9 North, is on the East Pacific Rise, a seafloor spreading center about 800 kilometers (500 miles) southwest of Acapulco, Mexico. Dawn describes it as "a very fast-spreading center," because it's spreading at a rate of around 11 centimeters (4.3 inches) per year—about "the same rate that your fingernails grow, so that's very fast in terms of earth's geologic processes." Because the area is spreading so fast, it's volcanically active, resulting in hydrothermal vents (underwater hot springs) that form when the crust stretches and cracks and superheated fluids burst from the crust. These chemical- and mineral-laden fluids can reach mind-boggling temperatures, at times over 400°C (750°F).

As Dawn describes it, 9 North is "a really strange alien world," an underwater landscape of towering lava pillars, huge cracks in the seabed, and vents spewing superheated chemical soup, an environment seemingly too hot and toxic to support life. But it does support life—tubeworms, mussels, crabs, clams, fish, gastropods, and crustaceans cluster around these vents, living in complete darkness and surviving through chemosynthesis (a process in which organisms make food from chemicals rather than sunlight).

Among the vent-dwellers identified at 9 North before the eruption were masses of tubeworms, notably giant tubeworms (*Riftia pachyptila*), which can grow up to 2 meters (6.5 feet) in length and depend on bacteria living inside them to synthesize their food. Their neighbors on the vent included the smaller Pompeii tubeworms (*Alvinella pompejana*), one of the most heat-tolerant species ever discovered.

When, by chance, oceanographers Rachel Haymon and Karen Van Damm dived to the scene in *Alvin* shortly after the volcanic eruption, not much was left of the tubeworms (or the mussels and crabs, for that matter).

Giant tubeworms (*Riftia pachyptila*) at a hydrothermal vent on the East Pacific Rise.

The seafloor was littered with dead and dying creatures, some encased in lava, others torn apart.

Dawn explains that "the tissues of the dead tubeworms were charred and shredded," as if they had "exploded from internal heating and expansion of their bodily fluids." When the tubeworm remains were sampled and brought back up to the support ship, she says, "It really did smell of cooked meat from the flesh of these tubeworms, so that's where the name 'Tubeworm Barbecue' comes from."

Although the story did not end well for the tubeworms, the eruption of 1991 gave marine biologists a rare opportunity to study the resurgence of life in an environment devastated by geologic activity. Scientists such as Dr. Vicki Ferrini have been returning to 9 North at regular intervals, conducting time-series analyses of all the activities, biological and otherwise, in that part of the deep ocean.

It did not take long after the eruption—a matter of days—for microbial mats (multilayered sheets of microorganisms) to show up in the cracks of the new ocean floor. As Dawn describes it, because of the vigorous hydrothermal venting from those cracks, there was now "white gelatinous

Tubeworms destroyed by a volcanic eruption on the East Pacific Rise, 1991.

bacterial material erupting from the seafloor, almost like a snow blower," creating a blizzard of white particles that coated everything.

These microorganisms proliferated after the eruption, without predators—temporarily. But a year later, when Dawn's team returned, the microbial mats had disappeared from the seafloor. What happened? The answer, it turned out, was the brachyuran crabs, which had returned as the vent system reset and the vent communities came back. These brachyuran crabs were feasting on the bacteria, and, Dawn says, "It is really something to see these crabs just scarfing the bacteria from the seafloor."

A "great time to be mapping"

Collecting seabed data from specially outfitted research vessels, such as the *Pressure Drop*, is prohibitively expensive, in terms of resources, and prohibitively slow, in terms of scale. Fortunately, when it comes to ocean mapping, we're entering an era of rapid technological innovation. "It's a great time to be mapping," Dawn says, pointing to several technological developments that could speed up the process of mapping the seabed and make it less costly, more precise, more environmentally friendly, and, ideally, more equitable.

Although sonar from ships is the most familiar method for collecting bathymetric data, ocean mappers can now collect and process data from an array of sources—not only at the surface (from ships), but above the surface and in the deep.

Satellites. Ironically, much of our current knowledge of the deep comes from on high. In the 1990s, scientists realized that satellites were observing and recording gravity-induced fluctuations in the surface of the ocean. Although to our eyes the ocean looks mostly flat, it's not—the effects of gravity create subtle bulges and dips on the surface, mimicking the shape of the seafloor below. A radar altimeter on a satellite can map these fluctuations, which can then be used to estimate ocean depth.

The problem is that, though satellite altimetry gives us a general view of the ocean floor, the measurements it returns aren't usually accurate enough, or high-resolution enough, for scientific purposes or for exploration. What altimetry gives us is only an approximation—but for vast swaths of the ocean, that's all we have.

> *We've got satellite altimetry, or a giant blank map. … I really think we should just remove satellite altimetry from the maps. It gives people a false sense of completion.* —Cassie Bongiovanni

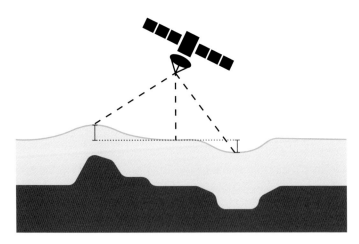

Altimeters on satellites can detect small differences in sea surface height as an indication of changes in Earth's gravitational field, which in turn are estimates of ocean depths around the world.

Bathymetric lidar. Unlike satellite altimetry, bathymetric lidar can produce incredibly high-resolution maps of the ocean floor. Lidar—an acronym for "light detection and ranging"—is a remote sensing method that uses laser pulses to measure the distance between objects. Bathymetric lidar uses pulses of high-intensity green light to collect information about the seabed, which, along with other data, generates detailed 3D maps—much more detailed, even, than those collected by multibeam sonar. Next-generation lidar technology will be more compact and require less power, allowing it to be deployed more widely, on submersibles or remotely operated vehicles.

Remotely operated vehicles (ROVs). Remotely operated vehicles allow us to explore the ocean without dipping so much as a toe into the water. These underwater robotic machines—which can be as small as a toaster or as large as an SUV—are uncrewed and controlled from a surface vessel, to which the ROV is tethered. ROVs can be equipped with cameras, lights, sonar, and even lasers, sending data

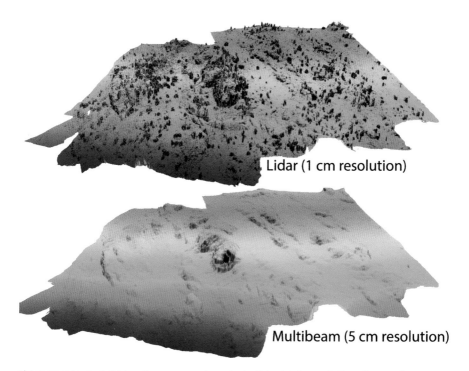

Lidar uses laser light to allow researchers to build a high-resolution, three-dimensional model of the deep seafloor (top) that is more detailed than bathymetric maps collected by traditional multibeam sonar (bottom).

The scientific lander *Flere* on the ocean floor.

and images back to the ship via the tether (a bundle of cables); some also have manipulator arms and other instruments to collect samples from the deep.

As NOAA explains, "ROV operations are simpler and safer to conduct than any type of occupied-submersible or diving operation because operators can stay safe (and dry!) on ship decks. ROVs allow us to investigate areas that are too deep for humans to safely dive themselves, and ROVs can stay underwater much longer than a human diver," allowing for more extensive surveys. In other words, these diligent machines don't need bathroom breaks, and they don't get the bends, so they can spend more time scurrying around the seafloor.

Autonomous underwater vehicles (AUVs). The difference between AUVs and ROVs is that AUVs, as the name implies, are autonomous: they're untethered and do their job independently, without a physical connection to a human operator. Like *Flere*, *Skaff*, and *Closp*, the landers that supported the *Limiting Factor*, AUVs are programmed with instructions, and then off they go, into the deep, collecting data. AUVs may be equipped with an array of instruments—Caladan's landers, for instance, were fitted with baited traps and high-definition cameras, which yielded significant scientific findings. When their tasks are completed, if all goes well, the AUVs return to a preprogrammed location. (Sometimes, like *Flere*, they get stuck, or go rogue.)

Today, ocean scientists are exploring the use of long-range AUVs for deep-ocean missions lasting weeks or months. These battery-powered vehicles can be launched and controlled from shore, rather than from ships, and are not only cost-effective but emissions free—a win for both science and the environment. Some can even be deployed and retrieved by uncrewed surface vehicles, also known as ocean drones or drone ships.

In Memoriam: *ABE*

A 1995 *Alvin* expedition served as the sea trial for an amazing new robot at the time, named the Autonomous Benthic Explorer, or *ABE*. If you look closely at the design of the vehicle, especially on the side, you'll see NCC 1701 B, which actually means the *Starship Enterprise* second version for undersea. This is because the engineers who designed and built it (Al Bradley and Dana Yoerger) were amazing *Star Trek* fans. They were hilarious to go to sea with, and it turns out that the design of this vehicle to make it look like the *Starship Enterprise* was extremely effective! I thought it was an engineering stroke of genius.

And in its heyday, *ABE* was really a wonderful partner to *Alvin* because he could explore on a preprogrammed track all night long while *Alvin*'s batteries were charging on the ship. And *ABE* could conduct

ABE (Autonomous Benthic Explorer), developed by Woods Hole Oceanographic Institution in 1994, with a design resembling the *Starship Enterprise* from the *Star Trek* science fiction series.

super-high-resolution bathymetric surveys, could measure the magnetic signals from the seafloor as well as the salinity and temperature in the deep waters. And so, the normal routine for *ABE* was to do its surveys at night, then surface in the morning. We'd get the data from *ABE*, and we were able to give some of the observations from *ABE* to the next set of *Alvin* divers to help them with their observations and their dive plan for the day. And the bathymetry from *ABE* was amazing, thanks in large part to Dr. Vicki Ferrini's work many years ago as she worked on the microbathymetry coming from *ABE*.

Now, very sadly, *ABE* was lost at sea on March 5, 2010, off the coast of Chile, but with a smile on his face and in his electronics to the very end. So, rest in peace, *ABE*.

Uncrewed surface vehicles (USVs). Also known as sea drones, USVs come in different shapes and sizes, with a wide range of sensors, and are used for a variety of purposes, from environmental monitoring to mapping to military missions. Some are powered by sail or solar, making them much more economical and efficient than crewed survey ships. Using these drones for mapping and research also reduces fuel consumption, carbon emissions, and noise pollution at sea.

With all this automation, oceanographers like Dawn are dreaming of a future where autonomous robots gather data 24 hours a day, 365 days a year, without needing a single ship to launch them or a single human to babysit them. If we rely on traditional ship-based methods, it could take us centuries to map the entire ocean. But new technologies such as AUVs and USVs could significantly speed up this process. Perhaps, in the future, there might be fleets of them, working together, sweeping the oceans with their sophisticated sensors and sending back "tons and tons and tons of beautiful data" to make beautiful maps.

Apart from efficiency, automation offers another key advantage, according to Dawn. It could make ocean mapping accessible to those who've historically been excluded from sounding the deep. As Dawn points out, "Following in the footsteps of Marie Tharp (who did finally go to sea herself), more and more women are going to sea today, and they come from all reaches of the planet, including the so-called Small Island Developing States (SIDS), also known as the Big Ocean or Blue Continent States. And if someone learns to operate an ocean drone, all they really need is an internet connection, anywhere, at sea or onshore, to participate in mapping."

Expanding access is precisely the mission of organizations such as Map the Gaps (for which Dawn was a flag-bearer on her dive), a nonprofit "committed to growing awareness, increasing diversity, and promoting equity in ocean mapping." Its goal is to "map the gaps" in seafloor data so that conservation organizations, communities, explorers, and researchers are no longer in the dark about the deep. Acknowledging the immensity of the task, the group advocates for its urgency: "We can't protect what we don't understand and we can't understand what we don't measure."

One way that Map the Gaps and other organizations aim to achieve this goal is by creating a global community of academics, explorers, and citizen scientists to crowdsource the data that's missing from the map.

Crowdsourced bathymetry (CSB). Dawn explains that "there's the expression *cheap and deep*, meaning you don't have to have a huge, sophisticated research vessel or a $20 million robot to map the ocean floor. Everybody can contribute." All kinds of oceangoing vessels— from cruise ships and cargo vessels to commercial fishing boats, from superyachts to weekend pleasure craft—can contribute depth readings as they go about their usual business. These vessels-of-opportunity, as they're called, can collect data from standard navigational instruments or from data loggers supplied by organizations such as Map the Gaps. The results, once collected, processed, verified, and standardized—a fairly labor-intensive process—can be used to fill in some of those blank spots on our bathymetric maps.

In addition, Dawn notes, "We're working with various industries that have bathymetric data that they have not released yet." National governments and industries such as oil, mining, and surveying have amassed stockpiles of proprietary data over the decades. If these organizations could be prevailed upon to make this data public (even at a lower resolution to protect their interests), it would speed up ocean mapping enormously. And that would benefit us all.

GIS. No matter how much data you have, it's just that—data. It doesn't speak for itself; it doesn't tell a story; it's not something you can use, query, or share with others. That's where GIS applications come in, with their power to process, manage, analyze, and map massive amounts of spatial data. Without GIS and its specialized bathymetry applications, mappers like Cassie Bongiovanni would be drowning in data—or laboriously hand-plotting profiles of the seafloor like Marie Tharp. Modern bathymetry, and the ability of oceanographers across the globe to share and interpret their data in real time, depends on GIS.

Mapping a deep-sea coral reef

In 2024, scientists from NOAA and the nonprofit Ocean Exploration Trust announced that they'd mapped the largest coral reef yet discovered in the deep ocean off the coast of the United States. Though researchers had known for decades that there was coral off the Atlantic coast, they didn't know how much. But thanks to state-of-the-art underwater mapping technology, these scientists were able to generate 3D images of the ocean floor and determine the reef's size. And it's huge: an area almost three times the size of Yellowstone National Park, extending from Florida to South Carolina. "It's eye-opening—it's breathtaking in scale," said Dr. Stuart Sandin, a marine biologist at the Scripps Institution of Oceanography. The reef lies in the twilight zone of the ocean (200–1,000 meters/655–3,280 feet), where sunlight doesn't penetrate. So, unlike tropical coral reefs, which rely on photosynthesis, deep-sea coral filters biological particles from the water for energy. In the words of Dr. Derek Sowers from the Ocean Exploration Trust, this enormous reef "has been right under our noses, waiting to be discovered." What other "breathtaking" discoveries await researchers as they map the remaining 75 percent of our ocean with GIS?

Mapping the entire ocean floor is a daunting task. There's still a long way to go. But with the concerted efforts of researchers, governments, industries, and communities around the globe, it can be done. And there's still so much to discover in the deep. As Dawn notes, "Ocean science is one of the few fields in which major groundbreaking discoveries can actually become the norm. There is so much left to explore that every seagoing expedition has a really good chance of making a new discovery." Looking to the future, the eminent oceanographer Dr. Robert Ballard has said: "The generation of kids in middle schools right now will explore more of Earth than all previous generations combined. Let that sink in. They're going to explore more of Earth than everyone that's been on this planet before them."

That's an exciting prospect and an inspiring one, but oceanographers like Dawn Wright and Sylvia Earle know that we need to act now; the clock is ticking. As Dawn was rudely reminded when she dived to the deepest spot on Earth and found a beer bottle dumped there, our oceans are in trouble.

Chapter 8

DIVING DEEPER

" *It turns out that the ocean is not too big to fail, unfortunately. The good news is that it's also not too big to fix.* —DAWN WRIGHT (as inspired by The Ocean Panel)

When people ask Dawn to describe her experience at Challenger Deep—not the scientific accomplishment but her personal, subjective impressions—she often has a hard time putting it into words. How to explain what that dive meant to her, not only as a scientist but as a human being?

But one day, in an interview with Parley for the Oceans, Dawn found the words. She began by describing what she and Victor saw down there: "We saw tiny creatures that live down there, such as the anemones and the sea cucumbers, the little amphipods, these little creatures that can withstand the 16,000 pounds per square inch of pressure, living in complete darkness." Then she added: "Those small, seemingly insignificant creatures, along with insignificant me and my colleague Victor making it all possible, are part of our little community. We are part of it too, we are part of the totality of life on this planet and—pivoting to broader thoughts about life on the planet and the miracle that is life on this planet—we all matter, and we're all interconnected." This idea of the value and interconnectedness of all life is a driving force in Dawn's work as an oceanographer, an educator, and a voice for a more humane and inclusive practice of science.

In another interview, Dawn reflected further on the mystery of life in the deep: "I don't know why these creatures exist," she said. "They live in complete darkness, in cold temperatures, but that is their world, it's part of our world. ... And then when those of us who are scientists have a chance to see these worlds, these worlds within a world that no one else gets to see, let alone study, it just reinforces for me how amazing and precious it all is."

From a planetary perspective, creatures (like us) who live on dry land and breathe the air are anomalies: most life on Earth exists underwater, much of it in the cold and dark of the deep ocean. "Our everyday world is not typical of life on Earth; if anything, it's the surface environment that's strange," writes author Riley Black. "Most of the planet is made up of deep-ocean habitats."

Nobody knows for sure what percentage of all life on Earth exists in the ocean, as most of the species that live in the deep have never been seen by humans and are still undescribed by science. "We live on an alien planet," Black writes. "We barely know our home planet at all." Estimates range wildly, but reputable sources, such as the UN, cite a range of 50–80 percent: 50–80 percent of all life-forms on this planet are believed to be quietly existing beneath the surface, from the sunlight zone to the hadal depths, in conditions that, by human standards, are extreme and unsurvivable. In fact, astrobiologists, who study the possibilities of extraterrestrial life, are studying these deep-sea creatures for clues about how organisms on other, harsher planets might have evolved.

And yet we know so little about what's down there. Almost every deep-sea scientific expedition discovers something new: a previously unidentified species, a species at previously unrecorded depth, an entire habitat teeming with life, a creature unlike any other. Even under the crushing pressure of full ocean depth, Dawn's dive and others like it have recorded a variety of life-forms adapted to the abyss: giant amphipods (shrimplike crustaceans), holothurians (sea cucumbers), xenophyophores (possibly the largest single-celled organism on the planet), decapod crustaceans, tubed anemones, and hadal jellyfish, to name a few.

> " *The thing that I love about the deep ocean is that every time you dive, every single time, you see something you've never seen. And every once in a while, you're going to see something that* nobody has ever seen. —James Cameron, director

In 2023, Dr. Alan Jamieson (chief scientist from the Five Deeps expedition) and his colleagues identified a new species of snailfish, the deepest fish ever recorded, at a depth of 8,336 meters (27,349 feet) in the Izu-Ogasawara Trench, southeast of Japan. Unlike the toothy, gnarly monster you might picture lurking in the deep, the snailfish is small, pink, and translucent, with delicate fins and no scales—not a monster but a svelte survivor.

Hadal jellyfish (*Pectis profundicola*) at 7,396 meters (24,265 feet), as recorded by the *Flere* hadal lander that accompanied Victor and Dawn on their dive to Challenger Deep. Snailfish to the right enjoying a snack, with lander arm in the foreground.

As a species, we expend enormous amounts of energy, resources, and imagination searching for aliens on other planets (and why not?) while paying scant attention to the aliens that have been here all along—on our own planet, in the deep. If we want to find beings whose biology is mind-bogglingly different from our own, we don't need to ping other planets. They're right here, creatures stranger and more otherworldly, to human eyes, than any extraterrestrial Hollywood could dream up. As Dawn discovered, to witness these (literally) outlandish creatures at home in their underwater world is to experience awe.

And there are millions of them—myriad life-forms, with their unique adaptations and survival mechanisms, that scientists believe evolved from single-celled microbes living near hydrothermal vents at least 3.5 billion years ago. According to Ocean Census, a global alliance working to discover and protect ocean life, 75–90 percent of the estimated 1 million to 2 million marine species remain undiscovered. The ocean is the largest ecosystem on Earth, and the largest remaining area of wilderness: we urgently need to protect it, but as conservationists remind us, we can't protect what we don't even know.

The ocean is also home to an astounding phenomenon that most of us never witness: the largest daily migration on Earth. Every day, all over the world, billions and billions of creatures that live in the deep sea migrate up and down the water column; they spend the day hiding from predators in the depths and then migrate en masse to the surface at night. These small marine organisms, known as zooplankton, feed on phytoplankton in the surface waters, so they travel as far as 1,000 meters (3,280 feet) "uptown" in the ocean to dine—and to avoid being dinner (for predators that depend on light). Adding to the biomass of this migration, other species of fish accompany them, to prey on the zooplankton and to avoid being preyed upon, in turn, by bigger fish.

We're talking an estimated 10 billion tons of biomass on the move, mostly out of sight, every single day. In addition to being an extreme endeavor—for a tiny organism, it's the equivalent, in human terms, of running a speedy 10K before and after dinner—this daily mass migration plays an essential role in carbon sequestration. Carbon ingested at the surface is excreted on the ocean floor, where it can remain for thousands of years, locked away.

As we know from Dawn and her fellow scientists, the health of ocean ecosystems is crucial to the health of the planet (including us), and biodiversity is crucial to the health of an ecosystem. But we are only just beginning to understand the nature and extent of biodiversity in the ocean. A 2022 study by researchers from the Natural History Museum in London found that about 60 percent of the DNA extracted from deep-ocean sediment couldn't be identified as animal, plant, or bacteria. This unclassifiable DNA, the researchers suggest, could represent "entire lineages of marine life that have yet to be described," an exciting challenge for marine biologists seeking to understand the past and present forms of ocean life—and possibly the origins of all life on Earth.

Today we have deep-sea technology, such as ROVs and AUVs, that enables us to reach and study areas of the ocean that were inaccessible before. Yet, at the same time, extinctions from habitat loss and climate change are escalating exponentially, putting an estimated 20 percent of marine species at risk. Will there be time for ocean scientists to identify and protect all these deep-sea organisms, all these strange and wonderful creatures representing eons of evolution and adaptation, before it's too late? Before human-driven threats such as pollution, climate change, and seabed mining alter the ocean forever?

Weird names for weird creatures

The common names of deep-sea creatures are their own kind of poetry—names that are beautiful and dreamlike for some creatures, dark and menacing for others, and just plain rude for the rest. (*Blobfish? Sea pig?* Really?) Here's a sampling of some creative nomenclature from the deep:

- Brittle star (*Ophiothrix fragilis*)
- Harp sponge (*Chondrocladia lyra*)
- Sea angel (*Gymnosomata*)
- Cosmic jellyfish (?—might be a new species)

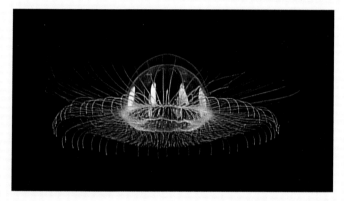

Cosmic jellyfish, spotted floating in the ocean depths near American Samoa on February 21, 2017.

- Elvis worm (*Peinaleopolynoe*)
- Vampire squid (*Vampyroteuthis infernalis*)
- Gulper eel (*Eurypharynx pelecanoides*)
- Black swallower (*Chiasmodon niger*)
- Zombie worm (*Osedax*)
- Yeti crab (*Kiwa hirsuta*)
- Dumbo octopus (*Grimpoteuthis*)
- Cookie-cutter shark (*Isistius brasiliensis*)
- Cockeyed squid (*Histioteuthis heteropsis*)
- Flabby whalefish (*Gyrinomimus grahami*)
- Faceless fish (*Typhlonus nasus*)
- Headless chicken monster (*Enypniastes eximia*)

The peanut butter connection

You probably don't think about the ocean while making a peanut butter sandwich (why would you?), but, as Dawn liked to remind her students, it's a substance from the ocean that makes your peanut butter spreadable, your toothpaste squeezable, and your soy milk drinkable. That substance is carrageenan, a compound extracted from red algae (seaweed) and mainly used as a stabilizer for foods—take a look at the ingredients of your favorite pudding or ice cream sometime. Agar, another substance extracted from red algae, is also widely used to stabilize cosmetics, thicken yogurt and jelly, and make the gelcaps for meds. Alginates, derived from brown algae (kelp), are used as thickeners and stabilizers in products ranging from cake mix to shaving cream. In fact, unless you live in a cave, it would be difficult to find personal-care products in your home that don't include ingredients from the sea.

Meanwhile, scientists are busy isolating compounds from marine organisms, such as algae, fungi, sponges, and bacteria, that they hope will help fight human diseases. Researchers have extracted at least 20,000 new biochemical substances from marine creatures to date, and dozens have reached clinical trials. These trials could lead to treatments for common and intractable ailments such as diabetes, cancer, arthritis, heart disease, and Alzheimer's disease.

These biopharmaceutical goals are not just pie in the sky. Two antiviral drugs already on the market—one to treat herpes infections, the other effective against the HIV virus, which causes AIDS—derive from compounds first isolated in the 1950s from a Caribbean sponge. More recent discoveries include ziconotide, used to treat chronic pain, which is derived from the toxins of a marine cone snail, and a promising new drug from a deep-sea microbe to fight glioblastoma, an aggressive (and, until now, untreatable) form of brain cancer. One compound from a deep-sea sponge also looks like a good candidate to attack the virulent, antibiotic-resistant superbug MRSA. And these are just a few examples from a red-hot area of research.

Dawn's principal area of study is hydrothermal vents, and she is fascinated by the adaptations of species that live in such extreme environments. "They are living with chemicals coming out of the seafloor that are like a toxic sewer because of the zinc," she says, "and the fluids coming out of those vents are toxic. And then those fluids are coming out at 400°C. So how are they able to withstand that? There are adaptations from these creatures that we can certainly use in biopharmaceuticals."

Scientists are particularly interested in sessile sea creatures—i.e., organisms that are fixed in place, like corals or sponges—because, think about it, how do these poor critters defend themselves? They can't scuttle away from predators, nor do they have teeth, claws, or often even a shell. The answer, it seems, is that they have powerful chemical and antimicrobial defenses, which researchers are studying for their possible disease-fighting benefits. Sessile organisms are plentiful around hydrothermal vents, and Dawn emphasizes how much we still have to find out about them: "With new species consistently being discovered, there may even be a cure for cancer or COVID in these places," she says. This is why, she adds, "it's important to study and to know as much as we can about our own planet."

Threats to our ocean

By now, we're probably all aware that our oceans are in trouble, facing what UN Secretary-General António Guterres called, in 2023, "the triple threats of climate change, biodiversity loss, and pollution." Many experts would also add overfishing to that list. And we may even be aware of another threat on the horizon: the push to begin mining the seabed for its minerals, regardless of the damage this may cause or the species that may be wiped out. That's the bad news.

The good news, according to Dawn, is that it's not too late to save the ocean, and that people across the globe, scientists and nonscientists alike, are working together to do just that. In 2017, the UN General Assembly declared 2021–2030 the UN Decade of Ocean Science for Sustainable Development, or the "Ocean Decade." Its vision is "The science we need for the ocean we want," and its goal is to foster collaboration across national and disciplinary boundaries to produce "transformative ocean science solutions" for a healthy, resilient, and sustainable ocean.

It's an ambitious initiative, but even so, we can't leave the fate of the ocean to the scientists and policymakers. As Dawn says, "The survival of our oceans, the survival of our one interconnected ocean, and the survival of our planet is dependent upon all of us." Dawn's focus, as an oceanographer, is on high-resolution mapping of the entire seabed. But each one of us, she reminds us, can make a difference, as part of "the family of humanity that is doing everything we can to save our ocean."

Plastic is forever

You've probably seen the heartbreaking images: the sea turtle with a plastic straw up its nose, the seal strangled by plastic netting, the seabird suffocating in a plastic bag. In many parts of the world, a stroll along a sandy beach means skirting a tideline composed entirely of plastic: bottles, toys, crates, bags, buoys, ropes, flip-flops, and who knows what else. It's disgusting. It's disheartening. And it's slowly killing our oceans.

Like the beer bottle that greeted Dawn and Victor in Challenger Deep, this type of pollution is perhaps the most visible evidence that, in Dawn's words, we are "trashing our only home," the watery Earth. More than 400 million metric tons of plastic are produced every year, much of it in the form of packaging, and, of that, at least 8 million tons end up in the ocean.

But it's not only the big stuff, the plastic bags and bottles, that we need to worry about. Potentially much more damaging are the microplastics: tiny pieces of plastic, no bigger than grains of sand, that have accumulated everywhere on Earth, in the ocean and on land, even in the Arctic. They, in turn, break down into even tinier particles that can enter our blood and our cells; scientists are still trying to determine their long-term effects on human health.

These minute particles result from the breakdown of larger plastic products over time and, thanks (or no thanks) to plastic's durability, last, essentially, forever. Some additives and chemicals in plastic are known to be toxic, and, as the plastic breaks down, it can absorb other toxins as well. When sea creatures, from plankton to whales, ingest these particles, they enter the food chain; microplastics have been found in the seafood we eat as well as in the water we drink. A 2024 study found that a liter of bottled water contains, on average, nearly a quarter of a million pieces of microplastic. It's not clear if, or how, they harm us, but surely the human body didn't evolve to absorb plastic. Not yet, anyway.

In 2020, Dr. Alan Jamieson and his team from Newcastle University identified a new species of amphipod in the Mariana Trench that they named—wait for it—*Eurythenes plasticus*. Why? Because this tiny creature, living at a depth of 6,096 meters (20,000 feet), had microplastics incorporated into its tissues. This is true

Trash in the ocean.

of every single specimen that's yet been found. As Dawn explains, "The genus is *Eurythenes* but the species is *plasticus* and that is tremendously tragic." Not only are plastics in our air and our water, but they're also in the bodies of organisms that live in the remotest, most inaccessible parts of the planet.

There are, however, glimmers of hope on the horizon. For starters, more than 90 percent of the plastic that pollutes our planet comes from single-use plastics, which we, as individuals, have the power to avoid. And, in February 2022, the UN Environment Assembly adopted a historic resolution to "develop an international legally binding instrument on plastic pollution, including in the marine environment." The UN aims to complete negotiations by the end of 2024 on a global treaty that would curb global plastic production and pollution. With so many nations involved, it's a fraught process that risks becoming deadlocked or hijacked by oil and gas interests. Environmentalists and ocean scientists are watching closely. In the words of Dr. Richard Thompson, who first identified microplastics, such a treaty represents a "once-in-a-planet opportunity"—but only if we get it right.

Mining the deep

" *How do we use the ocean without using it up?*
—Dr. Ayana Elizabeth Johnson, cofounder of the Urban Ocean Lab

They're dark brown, black, or reddish brown in color, with a rough, bumpy surface. They're typically potato-sized, though they can be as small as marbles or as big as footballs. They cover an estimated 70 percent of the seafloor, like cobblestones in some areas, and they've set off a 21st-century gold rush that may alter the ocean forever.

What are they? Known as manganese or polymetallic nodules, they're rock-like accumulations of nickel, cobalt, copper, manganese, and other elements, formed over millions of years as the minerals and metals in ocean water accrete around a hard object, such as a shark tooth or fish bone. These nodules take their time, building up at a rate of millimeters every million years, and they're an essential part of the ocean ecosystem, habitats for a diverse array of animals and microorganisms.

These deep-sea nodules were dredged up by HMS *Challenger* in the 1870s, but they've become the focus of mining interests around the world in recent decades, as demand for these metals has surged. These drab-looking rocks are rich in the metals we currently use to make batteries for our phones, laptops, and electric cars, as well as for "green" technology, such as wind turbines and solar panels. And that's why deep-sea miners are prospecting the abyssal plains, planning to scoop up these ancient nodules from the seabed with giant autonomous machines.

Mining companies have branded the nodules "a battery in a rock," claiming that seabed mining is more environmentally friendly, and less harmful to humans, than land-based mining and can provide most of the metals we need as we move away from fossil fuels. Conservationists and ocean scientists, on the other hand, have called for a moratorium on seabed mining, arguing that we don't yet know enough about the ocean floor to assess the potential consequences of ripping it up—both for the organisms that live there (biodiversity) and the carbon that's stored there (carbon sequestration). It could take decades, they say, to fully understand and weigh the risks of mining the deep.

"If this goes wrong, it could trigger a series of unintended consequences that messes with ocean stability, ultimately affecting life everywhere on earth," Pippa Howard, the director of Fauna and Flora International, told *Time* magazine. She

Polymetallic nodules, rich in minerals, cover the ocean floor.

notes that these nodules are part of a biome roughly the size of the Amazon rain-forest. "They've got living ecosystems on them. Taking those nodules and then using them to make batteries is like making cement out of coral reefs." As Sylvia Earle reminds us, it's not just rocks and water down there—it's alive, a living sys-tem. And once it's gone, we can't get it back.

Dawn, too, has expressed reservations: "We cannot, at this point, allow the seabed in those areas beyond our national jurisdiction to just be openly mined," she said in 2023. "We do not know what the consequences are yet. Ecosystems are not just the fish that swim at the surface or fish that swim in the waters beneath the surface, but the fish that live on the bottom and all the other creatures that live on the bottom—their habitats." She, too, compared this ocean wilderness to the Amazon rainforest, and said, "We can't just plow through and mine in the Amazon jungle and think that there are no consequences there. Similar thing in the ocean." Dawn also points out that battery manufacturers are already moving toward cobalt-free options, such as lithium-ion batteries, so demand for the met-als in the nodules may already have peaked.

The machines that are currently being designed and tested to collect the nod-ules from the seabed aren't nimble robots with tweezer-like arms; they're more like giant tanks, weighing about 25 metric tons each and vacuuming up a 10-cen-timeter (4-inch) layer of sediment—and all the life in it—as they plow over vast swaths of the ocean floor, grinding it to a slurry. In addition to the noise and light that would disrupt the deep, these machines stir up, and later pump down, plumes of sediment that could travel hundreds of miles in the ocean, damaging huge areas of habitat.

So who gets to decide whether seabed mining goes ahead? The International Seabed Authority (ISA), an intergovernmental body representing 168 countries, was created by the UN in 1994 to oversee mining in the high seas—those areas of the ocean that, according to the UN Convention on the Law of the Sea, are the "common heritage of humankind." In principle, then, beyond the exclusive economic zones that coastal nations control (up to 200 nautical miles from their shores), the ocean belongs to all of us, to humankind.

The ISA is still deliberating whether and how mining should be permitted to proceed, but meanwhile, it has issued exploration contracts to numerous nations, including China, Russia, and the UK, in the nodule-rich Clarion-Clipperton Fracture Zone. The CCZ, as it's known, is a deep abyssal plain between Hawaii and Mexico, about the size of Europe. Before the ISA grants any commercial licenses, it requires mining contractors to undertake three years of impact assessment—and, ironically, that's how scientists are discovering the rich biodiversity in the CCZ. On land, it's rare for scientists to discover a new species, an animal that hasn't been described by science. But researchers working in the CCZ report finding hundreds of new species on every expedition, and in every core sample and sediment they take.

With mining companies and countries such as the Pacific Island nation of Nauru pushing to accelerate the process, the ISA is aiming to finalize its regulations on seabed mining by July 2025. It's a contentious process, with more than 20 nations calling for a moratorium on deep-sea mining until its impact is better understood, and other nations eager to get industrial-scale mining under way. But, in a positive move, major companies such as Samsung, Google, and BMW have pledged not to use any materials sourced from the deep sea until the risks are fully assessed and alternatives, such as recycling, are exhausted.

In June 2023, Dawn was one of 804 marine science and policy experts from over 44 countries who signed a statement calling for a pause to deep-sea mining while scientists study its potential impact on the ocean environment. They believe that rushing ahead could have irreversible consequences. In the words of Dr. Diva Amon, a marine biologist who studies the CCZ, "By rushing in, we risk losing parts of the planet and species before we know them, and not just before we know them but before we understand them and before we value them."

Saving the high seas

The "high seas" is not just a term from pirate lore: in fact, the high seas are defined by international law as those areas of the ocean that lie outside the jurisdiction of any state. And in June 2023, after nearly 20 years of negotiation, 193 UN member countries finally adopted an international treaty to protect the high seas—a huge area encompassing about two-thirds of the global ocean. The UN High Seas Treaty is a legal framework that, once fully ratified, will establish policies to reverse biodiversity loss and ensure shared, sustainable development in the far reaches of the ocean. It's a significant milestone.

As Dawn explains, we can't allow unregulated activities—overfishing, dumping, mining, polluting, bioprospecting, and more—to continue in the high seas "as though it's a Wild West out there." Until this treaty comes into effect, the high seas will remain effectively lawless. This vast area of the ocean, 43 percent of the Earth's surface, doesn't "belong" to any nation. But, in a sense, it belongs to all of us, as the "common heritage of humankind."

Why should we care what happens way, way out there in the ocean? Apart from the obvious benefits of a healthy ocean, our marine wilderness needs to be protected and regulated, Dawn says, because the ocean is so dynamic. In the ocean, "everything is connected. Everything is in motion, everything is circulating." Any damage or degradation in one area of the ocean will eventually affect neighboring ecosystems, including our territorial waters. "Oceanographic science has known this for decades," Dawn points out, "and the idea of this High Seas Treaty is to agree to protect and regulate the activities of the high seas so that all of us can benefit."

Among other provisions, the treaty allows for the creation of marine protected areas—safe havens for vulnerable species of fish and plants. As Dawn notes, "Setting aside areas of the ocean to be preserved starts with choosing the right locations," areas that are hot spots of biodiversity. Networks of well-chosen marine protected areas would also benefit migratory species, such as whales and turtles. And choosing the right locations depends, to a large extent, on the insights we gain from modern digital mapping.

The climate challenge

❝ *Now it is our turn to take care of the ocean as it has taken care of us.*
—SHELBY O'NEIL, National Geographic Young Explorer

Ask Dawn (or pretty much any ocean scientist) what she sees as the gravest threat to our ocean, and she won't hesitate to say: climate change. As we continue to emit greenhouse gases, heating up the Earth's atmosphere, the oceans are heating up as well. Dawn notes that temperatures in the ocean are getting "alarmingly high," as the ocean absorbs around 90 percent of the excess heat that we humans generate on land. This, she says, affects not only our day-to-day weather patterns, as we've seen in recent years, but our long-term climate outlook: "Everywhere that you are on the planet, you are affected by this warming of the ocean."

"The other big consequence that we see," Dawn adds, "is the growing acidification of the ocean because as much as 25 percent of the greenhouse gases that are emitted on land are being absorbed. That's making the ocean more acidic, the coral reefs are in danger—and then there's the lack of oxygen. The ocean is losing its oxygen, especially in the deeper portions as well. Those are the big three—the temperature, the acidification, and the oxygen."

Deoxygenation, Dawn explains, is due to the disruption or destruction of tiny organisms in the ocean that produce oxygen through photosynthesis. And much of that destruction is due to industrial-scale commercial fishing—overfishing— that disrupts both the seabed and the water column with aggressive techniques such as trawling, netting, and dredging. Some scientists who study the effects of bottom trawling call it "marine deforestation" because of the irreversible harm it causes to ocean ecosystems, as well as the carbon that's released in the process.

Reflecting on her dive, Dawn noted that even the remotest spots in the ocean, such as Challenger Deep, are part of the complex climate system that keeps our planet habitable for humans:

> It struck me while being down in Challenger Deep how the oceans are heating up. The heat is circulated through these deepest trenches, all the way up to the surface of the ocean. Hence all of it matters, and all of it is buying us time in terms of climate change. The negative impacts of climate change would be hitting us so much more terribly, so much faster, if it weren't for the ocean absorbing a lot of it. It's keeping things at bay for as long as it can until we reach this terrible tipping point where even the ocean will not be able to help us.

It's a sobering assessment, but one shared by Kathy Sullivan, who observes that "these changes are driving the habitability of the planet into domains it has never been, not only in human life but in any phase of history that we can understand through the geological record."

For both Dawn and Kathy Sullivan, though, the gravity of the climate crisis is not a reason to despair; it's a reason to take action. And, for Dawn, climate action is also a form of social justice action. As marine biologist Ayana Elizabeth Johnson has noted, "People of color disproportionately bear climate impacts, from storms to heat waves to pollution"—and are the most concerned about the climate issue. Yet all life on Earth is interconnected, and, as Dawn says, "The quality of our natural world matters for every person and living organism on our planet, and where environmental impact is the greatest, so too is human suffering."

A call to action

Today, Dawn believes, we're in a period of awakening where issues of justice, equity, diversity, and inclusion are concerned. In ocean mapping as in other fields, this awakening is beginning to transform the practice of science and the ethics of conservation. In Dawn's view, it's no longer just a question of why we map the ocean and how, but also of who does the mapping and why it matters. It matters because, in Dawn's words, "Mapping can ultimately bring us all together, and that is certainly what we need to change the world."

As a keynote speaker at the 2023 Map the Gaps Symposium in Monaco, Dawn reflected on recent encouraging developments in her field, as it expands to include a wider range of individuals and communities who care about the ocean. "In mapping the ocean," she said, "what is now of paramount importance is not just the *where* and the *what*, but the *who* or by *whom*. And there are so many emerging programs that reflect this emphasis."

Here, in Dawn's words, are some examples:

- One of the most prominent and effective "empowerment" programs, in my view, is **Black in Marine Science**, which grew from a single tweet by its founder, Tiara Moore, into a full-fledged nonprofit, with nearly a dozen different programs for students, professionals, and the public, as well as a UN Ocean Decade Program. Similar organizations are emerging, such as Latinx in the Marine Sciences and Black Women in Ecology, Evolution, and Marine Science.

- The **Global Deep Sea Capacity Assessment team** of the Ocean Discovery League includes research assistants from countries all over the globe—emerging and empowered explorers from countries such as the Cook Islands, the Philippines, Fiji, Costa Rica, South Africa, and Morocco. This type of diversity used to be rare in ocean science but is now, thankfully, becoming more common. We at Esri are excited to be working with this organization.

- Nicole Yamase and her colleagues have launched the amazing **Madau Project** to reconnect Micronesian youth in Hawaii to their navigational heritage. If that's not a way of empowering interconnectedness and exploration, I don't know what is!

- **Fair Seas for All** is an initiative described in an important 2023 research paper that illuminates how, even though work at sea is essential to ocean science, transgender and gender-diverse scientists often face obstacles and harassment that make field research unwelcoming and even traumatic.

- On the research front, a project that Esri is proud to be a part of is a network of networks to implement the existing **Deep Ocean Observing Strategy** (**DOOS**). DOOS is a community-driven, international initiative fully endorsed by the UN Ocean Decade. Its goal is to align the deep ocean–observing community (including Seabed 2030 and Map the Gaps) in exploring collective solutions to the deep ocean challenges we face.

- Very, very important here, in terms of empowering a diverse cadre of explorers, is that DOOS is also funding six **Early Career Researchers**, who are not just participants but coleaders in each of the working groups. This is absolutely key to the project's success. And these six are the leading lights of a broader, diverse, cross-disciplinary cohort of more than 120 Early Career Researchers from around the world, called the **DOERs** (**Deep Ocean Early Career Researchers**). Two-thirds of the DOERs are outside the US, with 33 from developing countries or small island developing states.

- Another shining example is **WINGS WorldQuest**, based in New York City, the only organization in the world that awards completely unrestricted grants to women explorers of note—women of any race, culture, or age who are actively engaged in fieldwork that investigates questions about how the Earth works and have demonstrated the ability to think outside the box and follow an uncommon path (among other qualities). The grants are modest, but because they're unrestricted, they can be used for purposes such as childcare, care of an elderly parent, pet care, hiring an additional intern—anything that will further empower a woman explorer. WINGS also offers many other programs to elevate and build community around visionary women and their pioneering discoveries in science, exploration, and conservation.

These are hopeful initiatives in the fields of ocean science and exploration. But what if you're not an oceanographer or an explorer yourself? What if you failed high school science, live in a landlocked country, or don't like to get your feet wet—but you're still concerned about the health of the ocean? What, realistically, can you do? Should you just sit back, leave it to the experts, and hope for the best?

Absolutely not, says Dawn, who concludes with a heartfelt call to action, addressed to each one of us.

Dear Reader,

Even if you're not an ocean scientist, *you* can still follow these initiatives, and you can get involved yourself. What we're doing to the ocean right now (pollution, acidification, deoxygenation) has huge consequences for us all and for all the interconnected organisms on this planet. But each of us, no matter where we live, can take action to help protect the ocean and reverse some of the destructive trends that threaten it today. Because, on an ocean planet, the ocean surrounds, supports, and connects us all.

You might consider some of the following options:

- Take simple steps to recycle or greatly reduce plastic use, especially straws, since this affects the ocean no matter where you live. Plastic pollution of waterways is directly connected to declining oxygen levels in the ocean. Organizations such as the Algalita research institute, the Save the Albatross Coalition, and the Marine Conservation Institute all provide excellent tips and resources. You could even participate in (or organize) a beach or river cleanup!

- This summer, visit a "blue park" (a marine protected area) in person to gain a fuller appreciation of the life-giving resources the ocean provides—including oxygen!

- Join forces with Mission Blue and other similar organizations that showcase many of these "blue parks." At Mission Blue, anyone can nominate a section of the ocean to become a "blue park," also known as a Hope Spot, for protection and preservation.

- Use the Marine Protection Atlas to see current and emerging areas of protection.

- Use Esri's wide range of data and apps accessible in ArcGIS Living Atlas of the World to study the physical and biological happenings throughout the world's oceans.

- Follow Esri's ocean initiatives, including events such as the Esri Ocean, Weather, and Climate GIS Summit.

There's plenty that each of us can do to support marine conservation and help reverse the degradation of our world's ocean. And young people have a vital role to play in this—like the 12-year-old girl from Florida who reached out to me on social media a few years ago and ended up attending the International SeaKeepers Society event in 2022. Or the little boy from Albuquerque, New Mexico, who, after hearing about my dive to Challenger Deep, begged me to come and speak to his science club. Though I wasn't able to do that, I was able to reach out to his club and meet his club mates and his teachers, all Black, all eager to learn about my dive.

I'm so grateful that my Challenger Deep adventure has opened new ways for me to reach out to young people who, until now, had never seen themselves as connected to the ocean or ocean science—young people whose resilience and creativity inspire me with hope. In the words of National Geographic Young Explorer Shelby O'Neil, "Young people will take on the challenges of tomorrow, ranging from hard conversations to innovative conservation solutions, because even though we didn't create this mess, we are eager to clean it up." So much is happening today with the younger generations as agents of change.

As ocean scientists, we face the challenges of lack of funding, lack of attention and understanding by the public, partisan political divides, and getting our policymakers to prioritize ocean mapping, while also advancing conservation policies that center on equity and justice. Together, as humanity, we face the dangers and ravages of wars and climate change, among other existential threats. As we confront these challenges, we often hear the expression, "We're all in the same boat." But, given the inequities of our world, I find this sentiment by the Scottish writer Damian Barr much more apt: "We are not all in the same boat. But we are all in the same storm."

We're all in the same storm, but, as Tony Remengesau, former president of the Republic of Palau and a voyager to Challenger Deep, has pointed out, we all—all of humanity—need to be rowing in one direction, in our different boats, toward solutions to protect our ocean, our planet, and ourselves.

The good news is that we—a collective *we*—have the technology to do this. We also have centuries' worth of traditional ecological knowledge to draw on and the knowledge of Indigenous scientists, whose seafaring peoples have long been studying and protecting the ocean that sustains them. And together we must get to know every inch of the Earth, every crevice and fissure of our seafloor, to understand how our planet works. Only then can we fully understand the challenges we face and see the solutions clearly.

As another one of my heroes, Nainoa Thompson, a Polynesian master navigator, has said: "[As we] human beings are changing the Earth, it is now turning around and changing us. And we [often] don't know what to do, because we disconnected. The greatest thing we need to do is to become family of the Earth again, family of nature"—in other words, become stewards of the Earth by recognizing and acting upon our interconnectedness.

In my view, THIS is how we ultimately map the deep for the good of all.

Dawn J. Wright

—Dawn Wright

Acknowledgments

I t's not every day that valued colleagues at work come to you and ask if they can work with you on writing an entire book about the greatest adventure of your life, while also further educating the public about the crucial significance of mapping the ocean. So, for this, I must absolutely thank my Esri Press colleagues Catherine Ortiz, Stacy Krieg, Jenefer Shute, and Alycia Tornetta. This book was their idea! Jenefer's skill at researching my story and drawing out cogent insights, all the while wrapping it in engaging and compelling prose, was astounding. Many thanks to Alycia for her attention and skill in the overall editorial managing of the project. As always, it was a great pleasure to work with the entire Esri Press team! Thank you also to the incredible ArcGIS StoryMaps team, especially Will Hackney, who first brought my dive to life and helped provide the basis for this book.

As for the story itself, I must "deeply" thank my hero and friend Victor Vescovo and all his Caladan Oceanic team, as well as Rob McCallum and his EYOS Expeditions team. They made the Challenger Deep dive possible and opened an amazing new world for me and for others as a result. I must also acknowledge UCSB Professor of Earth Science Emerita Rachel Haymon, who gave me my first submersible dive as a graduate student and pushed me to conduct my science with a passion and precision that was unknown to me until I started working with her. My story is always and forever infused with the spirit of my mother, Jeanne Wright, and the sheer blessing, inspiration, and strength that she provided.

Cheers,
Dawn

Additional resources

Dive deeper at **mappingthedeep.com**.

Recommended reading

Baker, Aryn. 2021. "A Climate Solution Lies Deep Under the Ocean—but Accessing It Could Have Huge Environmental Costs." *Time*, September 10, 2021.

Cameron, James (dir.). 2005. *Aliens of the Deep* (documentary).

Black, Riley. 2023. *Deep Water*. Chicago: University of Chicago Press.

Casey, Susan. 2023. *The Underworld: Journeys to the Depths of the Ocean*. Doubleday.

Wright, Dawn. 2021. "The Courage to Step Out with Dawn Wright." *Kathy Sullivan Explores* podcast. September 23, 2021.

Wright, Dawn, and Sylvia Earle. 2018. "Deoxygenation of the Ocean Affects Everyone, So Act Now." *ArcNews*, Summer 2018.

Johnson, Ayana Elizabeth. 2020. "I'm a Black Climate Expert. Racism Derails Our Efforts to Save the Planet." *Washington Post*, June 3, 2020.

Sullivan, Kathy. 2020. "Why Exploration Could be the Key to Saving our Planet," with Dawn Wright. *Esri & The Science of Where* podcast, November 23, 2020.

Krajick, Kevin. 2004. "Medicine from the Sea." *Smithsonian Magazine*, April 30, 2004.

Scales, Helen. 2022. *The Brilliant Abyss: Exploring the Majestic Hidden Life of the Deep Ocean, and the Looming Threat That Imperils It*. Grove Atlantic.

Wright, Dawn. 2022. "To Save Earth's Climate, Map the Oceans." *ArcNews*, Winter 2022.

Trethewey, Laura. 2023. *The Deepest Map: The High-Stakes Race to Chart the World's Oceans*. Harper Wave.

National Geographic. 2021. "We're a Young Explorer and a Scientist, and These Are Our Ocean Stories. What's Yours?" *National Geographic* Education Blog, September 2, 2021.

Alberts, Ellizabeth Claire. 2022. "'We've Got to Help the Oceans to Help Us': Q&A with Deep-Sea Explorer Dawn Wright." *Mongabay*, August 2, 2022.

Young, Josh. 2022. *Expedition Deep Ocean: The First Descent to the Bottom of All Five of the World's Oceans*. Pegasus.

Organizations

Black in Marine Science, www.blackinmarinescience.org

The Deep Ocean Observing Strategy, www.deepoceanobserving.org

The International SeaKeepers Society, www.seakeepers.org

Map the Gaps, www.mapthegaps.org

Ocean Discovery League, www.oceandiscoveryleague.org

Seabed 2030, www.seabed2030.org

WINGS WorldQuest, www.wingsworldquest.org

Notes

Quotations and sources in the book are referenced here by page number.

Chapter 1

This chapter is adapted from "Challenge Accepted," by Esri's StoryMaps team, written by Will Hackney, November 8, 2022. https://storymaps.arcgis.com/stories /08a75d3687e749989840e7d236aab74d

3. *"It's exhilarating, and it never gets old:* "Dawn Wright: Rolling in the Deep," Parley for the Oceans, April 20, 2023. https://parley.tv/journal/dawn-wright-challenger-deep

3. *"That is part of the culture of Hawaii:* "Dawn Wright: Rolling in the Deep," Parley for the Oceans, April 20, 2023. https://parley.tv/journal/dawn-wright-challenger-deep

12. *"This is further evidence:* "Mission Accomplished: Photos from Dawn's Challenger Deep Expedition," by Victoria Phillips. Esri Industry Blogs, August 3, 2022. https://www.esri.com/en-us/industries/blog/articles/mission-accomplished-photos-from-the-challenger-deep-expedition

12. *"I don't know if I'll ever get to the moon:* "Dawn Wright: A Divine Abyss," *Biologos* podcast, May 18, 2023. https://biologos.org/podcast-episodes/dawn-wright-a-divine-abyss

14. *Flintstones' Quarry:* Dawn Wright, "Scichat: Deep Sea Adventures with Dr. Dawn Wright: A Pirate's Tale of Science & Exploration," *The Science Pawdcast*, September 20, 2023.

Chapter 2

18. *"I was eight when I decided:* "Dawn Wright," I Was a Kid website, by Karen Romano Young. https://www.iwasakid.com/dawn-wright

19. *"She and her siblings:* Dawn Wright, "The Courage to Step Out with Dawn Wright," *Kathy Sullivan Explores* podcast, September 23, 2021.

19. *"I don't think she ever saw: Kathy Sullivan Explores.*

19. *For a young Black woman to venture out:* Dawn Wright, transcript of interview with The HistoryMakers®, The Digital Repository for the Black Experience, ScienceMakers. November 27, 2012, p.16

19. *"I barely made it:* HistoryMakers, p. 12.

19. *"It was quite a shock": Kathy Sullivan Explores.*

20. *"So we went from Saskatchewan: Kathy Sullivan Explores.*

20. *Maui, in the 1960s:* HistoryMakers, p. 15.

20. *"I remember difficulty finding: Kathy Sullivan Explores.*

20. *A crossroads of the Pacific.:* HistoryMakers, p. 18.

20. *"Doing a little snorkeling: Kathy Sullivan Explores.*

20. *"Running on the beach:* I Was a Kid.

20. *"I was a typical kid: Kathy Sullivan Explores.*

20. *"My steady diet of science communication:* "Dawn Wright: Rolling in the Deep," Parley for the Oceans, April 20, 2023. https://parley.tv/journal/dawn-wright-challenger-deep

21. *"There were so many little kids:* "Dawn Wright: A Divine Abyss," *Biologos* podcast (edited), May 18, 2023. https://biologos.org/podcast-episodes/dawn-wright-a-divine-abyss

22. *"Read, read, READ:* I Was a Kid.

23. *"I realized the true power:* "We're a Young Explorer and a Scientist, and These Are Our Ocean Stories. What's Yours?" *National Geographic* Education Blog, September 2, 2021.

24. *"It was a very hard move:* HistoryMakers, p. 31.

24. *"I felt completely out of place:* I Was a Kid.

24. *"I had some really fabulous teachers: Kathy Sullivan Explores.*

24. *"I knew I was going to head off: Kathy Sullivan Explores.*

25. *"I wanted to be at a school: Kathy Sullivan Explores,* edited.

25. *"Texas A&M was not a place:* HistoryMakers, p. 44.

26. *"I reconsidered whether I was:* HistoryMakers, p. 45.

26. *"I had to sink or swim:* HistoryMakers, pp. 46–47.

26. *"I felt very deflated and distressed: Kathy Sullivan Explores.*

27. *"This is the story: Kathy Sullivan Explores.*

27. *"One of the most amazing:* HistoryMakers, p. 47.

27. *"So all of this was incredibly exciting:* HistoryMakers, p. 48.

28. *"Getting exposure to all of these:* HistoryMakers, p. 48.

28. *"There were 100 souls: Kathy Sullivan Explores.*

29. *"I've found that when you: Mapping the Deep* hub, "Meet Dawn." https://mappingthedeep-story.hub.arcgis.com/pages/meet-dawn.

29. *"When an iceberg would get:* Dawn Wright, "Deep Sea Drilling with Dawn," AGU's Third Pod from the Sun, November 1, 2018.

30. *"For a young person:* HistoryMakers, p. 49.

31. *"Was a very exciting field:* HistoryMakers, p. 51.

31. *"[Dr.] Ray Smith:* Dawn Wright interview, "Go with the Flow," University of California, Santa Barbara, Alumni Spotlights [no date]. https://www.alumni.ucsb.edu/stay-informed/alumni-spotlights/go-with-the-flow

31. *"To understand more: Kathy Sullivan Explores.*

32. *"It was absolutely thrilling:* Dawn Wright interview, "Go with the Flow," University of California, Santa Barbara, Alumni Spotlights.

32. *"It was just serendipity:* HistoryMakers, p. 54.

32. *"That was the first time:* HistoryMakers, p. 54.

34. *"We were so tired: Mapping the Deep* hub, "Meet Dawn."

34. *"I'd come very close:* HistoryMakers, p. 56.

34. *"We were able to look:* HistoryMakers, p. 56.

36. *The only Black female faculty member:* HistoryMakers, p. 67.

36. *"If I had limited myself:* HistoryMakers, p. 66.

36. *"We desperately need: Mapping the Deep* hub, "Meet Dawn."

36. *"Teaching is very hard:* HistoryMakers, p. 63.

38. *"The 800-pound gorilla:* Kathy Sullivan Explores.

38. *"I was sort of knocking:* Kathy Sullivan Explores.

38. *"Was looking for a bit:* Kathy Sullivan Explores.

38. *"I have not regretted:* "The Courage to Escape," by Dawn Wright. *Compass,* April 30, 2013. https://dusk.geo.orst.edu/compass.html

39. *"I always remind people:* Kathy Sullivan Explores.

40. *"A world where we don't:* "'Join the Rebel Alliance' – an *Openscapes* interview with Dr. Dawn Wright at MozFest," by Erin Robinson and Julie Lowndes. *Openscapes,* March 25, 2021. https://openscapes.org/blog/2021-03-25-rebel-alliance-dr-dawn-wright

40. *"I think my legacy:* HistoryMakers, p. 67.

Chapter 3

The first section of this chapter is adapted from the ArcGIS StoryMaps story "How Deep Is Challenger Deep?" by John Nelson, July 1, 2022. https://storymaps.arcgis.com/stories /0d389600f3464e3185a84c199f04e859

51. *"We concluded:* "On the Scientific Exploration of the Deep Sea," by Carpenter, Jeffreys, and Thomson. *Proceedings of the Royal Society of London,* vol. 18, 1869, p. 423.

51. *"The recklessness of beauty:* "On the Depths of the Sea," by Wyville Thomson. *The Journal of the Royal Dublin Society,* vol. 5, 1870, p. 322.

52. *"The window to a wholly new world:* Beebe, William. *Half Mile Down.* Harcourt, Brace, and Company, 1934, p. 65.

52. *"We were the first:* Beebe, p. 109.

54. *"When once it has been seen:* Beebe, p. 175.

Chapter 4

Sections of this chapter are adapted from the "Humanity of the Deep" ArcGIS StoryMaps story, by the Esri StoryMaps team, written by Will Hackney. https://storymaps.arcgis.com/collections /0655edef77e14d9caco3d147c10aa988?item=2

67. *"I have to explain:* Trethewey, Laura. *The Deepest Map: The High-Stakes Race to Chart the World's Oceans.* Harper Wave, 2023, p. 17.

67. *"I couldn't sleep:* Trethewey, p. 14.

Chapter 5

A section of this chapter is adapted from "Mission Accomplished: Photos from Dawn's Challenger Deep Expedition," by Victoria Phillips and Dawn Wright. *Esri Industry Blog,* August 3, 2022. https://www.esri.com/en-us/industries/blog/articles/mission-accomplished-photos-from-the-challenger-deep-expedition

75. *"We really should know:* From "Illuminating the Deep," ArcGIS StoryMaps story by Esri StoryMaps team, written by Will Hackney. https://storymaps.arcgis.com/collections/0655edef77e14d9caco3d147c10aa988?item=1

77. *"Get hooked on science:* Sigma Xi. "Women in STEM: 2015." https://www.sigmaxi.org /programs/critical-issues-in-science/diversity/women-in-stem/2015

78. *"A lust for adventure:* Great Big Story. "What It's Like to Walk in Space and Dive 7 Miles Below Sea Level." YouTube, August 17, 2020. https://www.youtube.com/watch?v=DO03IRwzvfc&t=2s

78. *"I loved the expeditionary part:* Neidl, Phoebe. 2021. "From the Ocean to Outer Space: The Adventures of Dr. Kathy Sullivan." *Rolling Stone*, March 3, 2021. https://www.rollingstone.com/culture/culture-features/kathy-sullivan-astronaut-challenger-deep-1132953

79. *"I was just ready to go:* Neidl, *Rolling Stone* interview.

79. *"A different kind of jigsaw puzzle:* Neidl, *Rolling Stone* interview.

80. *"Dr. Sullivan has been:* Neidl, *Rolling Stone* interview.

80. *"As a hybrid oceanographer:* EYOS Expeditions press release, June 7, 2020.

81. And conversely, going deep into the sea: Edited together from two sources: "Illuminating the Deep" ArcGIS StoryMaps story and youtube.com/watch?v=DO03IRwzvfc.

81. *"You're in a craft:* Neidl, *Rolling Stone* interview.

81. *"I've seen things:* Great Big Story, YouTube.

82. *"I have been immersed:* "Father's Day Tale: Kelly Walsh Dives Challenger Deep, 60 Years after His Father, Don Walsh, Became the First." *Eyos Expeditions* blog, September 28, 2023. https://www.eyos-expeditions.com/fathers-day-tale-kelly-walsh-dives-challenger-deep-60-years-after-his-father-don-walsh-became-the-first

83. *"The sub Limiting Factor:* WHOI press release, "WHOI Researcher Dives to Challenger Deep," June 26, 2020. https://www.whoi.edu/press-room/news-release/whoi-researcher-dives-to-challenger-deep

83. *"We were always surrounded:* "Nicole Yamase: Honoring Her Culture through Science," by Kate Uesugi. TheHumanist.com, May 26, 2021. https://thehumanist.com/features/articles/nicole-yamase-honoring-her-culture-through-science/

83. *"Our ancestors were scientists:* Letman, Jon. 2021. "Micronesian Scientist Becomes First Pacific Islander to Reach Ocean's Deepest Point." *The Guardian*, April 3, 2021. https://www.theguardian.com/world/2021/apr/04/micronesian-scientist-becomes-first-pacific-islander-to-reach-oceans-deepest-point

83. *"We belong all the way:* Letman, *The Guardian*.

84. *"I couldn't believe my eyes:* University of Hawaii. 2021. "1st Pacific Islander to Reach Ocean's Deepest Point Is UH Grad Student." University of Hawaii News, April 6, 2021. https://www.hawaii.edu/news/2021/04/06/ocean-deepest-point-grad-student

84. *"But to Pacific Islanders:* Macumber, Bobby, and Dan Smith. 2023. "Challenger Deep Is a Terrifying 11 km Underwater — but for Nicole Yamase, It Was Like 'Getting a Hug from the Ocean.'" For Stories from the Pacific, ABC Australia, August 11, 2023. Reproduced by permission of the Australian Broadcasting Corporation—Library Sales. © 2023 ABC. https://www.abc.net.au/pacific/nicole-yamase-dive-to-challenger-deep/102682288

84. *"It was almost spiritual:* US National Science Foundation. 2021. "#NSFStories: Voyage to the Bottom of the Sea," by Chris Parsons. June 9, 2021. https://new.nsf.gov/science-matters/nsfstories-voyage-bottom-sea

85. *"A pipe dream:* Alberts, Elizabeth Claire. 2022. "'We've Got to Help the Oceans to Help Us': Q&A with Deep-Sea Explorer Dawn Wright." *Mongabay*, August 2, 2022. https://news.mongabay.com/2022/08/weve-got-to-help-the-oceans-to-help-us-qa-with-deep-sea-explorer-dawn-wright

85. *"Victor has asked me:* Alberts, *Mongabay*.

86.　*"We have all kinds:* Wright, Dawn. 2023. ©2023 National Public Radio Inc. Excerpts from "Why Mapping the Entire Seafloor Is a Daunting Task, but Key to Improving Human Life" were originally published on NPR's *Short Wave* podcast on November 6, 2023, and are used with the permission of NPR. Any unauthorized duplication is strictly prohibited. https://www.npr.org/2023/11/06/1198908501/mariana-trench-challenger-deep-ocean-map-tsunami

88.　*"You're going in a machine:* Macumber and Smith, ABC Australia.

90.　*"We'd do eight-hour shifts:* Barford, Vanessa. 2013. "*Pisces III* Submersible: A Dramatic Underwater Rescue." *BBC News,* August 30, 2013. https://www.bbc.com/news/magazine-23862359

93.　*"I had come across this business anomaly:* "A Deep Dive into Plans to Take Tourists to the Titanic," by Tony Perrottet. *Smithsonian Magazine,* June 2019. https://www.smithsonianmag.com/innovation/worlds-first-deep-diving-submarine-plans-tourists-see-titanic-180972179.

95.　*"Consistent with the catastrophic loss:* US Coast Guard press conference, June 22, 2023. https://www.dvidshub.net/video/888022/coast-guard-holds-press-briefing-about-discovery-debris-belonging-21-ft-submersible-titan.

96.　*"Defining feature of the Limiting Factor:* The Five Deeps Expedition website: https://fivedeeps.com/home/technology/sub

98.　*"Extraordinarily brave:* Tsangaris, Kyriakos, and Johann Meyer. 2022. "Deepest Shipwreck Discovery: WWII Wreck of USS *Samuel B. Roberts* Found in Philippine Sea." *Sea Technology,* December 2022. www.sea-technology.com

106.　*"Symmetry and success:* Wright, Dawn. 2021. "The Courage to Step Out with Dawn Wright." Kathy Sullivan Explores podcast, September 23, 2021.

106.　*"A wonderful analogy to open science: Mapping the Deep* hub, "Meet Dawn." https://mappingthedeep-story.hub.arcgis.com/pages/meet-dawn

Chapter 6

A section of this chapter is based on the ArcGIS StoryMaps story "The Data behind the Search for MH370." *Geoscience Australia,* July 19, 2017. https://geoscience-au.maps.arcgis.com/apps /Cascade/index.html?appid=038a72439bfa4d28b3dde81cc6ff3214 and "Downed Airliner Search Mission Yields Map with Lasting Purpose," by Dawn Wright. *Esri Blog,* March 21, 2018. https://www.esri.com/about/newsroom/blog/mh370-search-mission-yields-map-with-lasting-purpose

110.　*"We need to act now:* Wright, Dawn, and Sylvia Earle. 2018. "Deoxygenation of the Ocean Affects Everyone, So Act Now." *ArcNews,* Summer 2018. https://www.esri.com/about /newsroom/arcnews/deoxygenation-of-the-ocean-affects-everyone-so-act-now

110.　*"The most exciting:* Wright, NPR *Short Wave* interview.

110.　*"Tsunamis come about:* Wright, NPR *Short Wave* interview.

111.　*"Have killed hundreds of thousands:* NOAA. 2017. "Tsunamis: What Can the Ocean Floor Tell Us about the Next Disaster?" March 7, 2017. https://www.noaa.gov/stories/tsunamis-what-can-ocean-floor-tell-us-about-next-disaster

112.　*"Another very, very important reason:* Wright, NPR *Short Wave* interview.

112. *"How do we ensure:* Keynote address to the 2023 Esri User Conference, Dr. Richard W. Spinrad, "The New Blue Economy & Predictability." https://www.esri.com/about/newsroom/arcuser/economy-predictability

112. *"Our maps show that 99 percent:* Wright, NPR *Short Wave* interview.

114. *"To monitor and protect:* Wright, NPR *Short Wave* interview.

115. *The global mean sea level:* NOAA/climate.gov. 2022. "Climate Change: Global Sea Level," April 19, 2022. https://www.climate.gov/news-features/understanding-climate/climate-change-global-sea-level

116. *"If we have additional accidents:* Wright, NPR *Short Wave* interview.

119. *"We are going to be putting:* Wright, NPR *Short Wave* interview.

119. *Climate change is the most basic:* Wright, Dawn. 2022. "To Save Earth's Climate, Map the Oceans." *ArcNews*, Winter 2022.

Chapter 7

121. *Gathering data from acoustic instruments:* Wright, Dawn. 2022. Adapted from "An Inspiring Journey to Map the Deepest Part of the Ocean." *ArcNews*, Fall 2022.

122. *Ewing had a hard time:* "Connect the Dots: Mapping the Seafloor and Discovering the Mid-Ocean Ridge," by Marie Tharp. *Lamont-Doherty Earth Observatory of Columbia: Twelve Perspectives on the First Fifty Years 1949–1999*, edited by Laurence Lippsett. Lamont-Doherty Earth Observatory of Columbia University, 1999.

123. *"It was a once-in-a-lifetime:* "Connect the Dots," by Marie Tharp.

126. *"We [Tharp and Heezen] were upsetting:* Nelson, Valerie J. 2006. "Marie Tharp, 86; Pioneering Maps Altered Views on Seafloor Geology." *Los Angeles Times*, September 4, 2006. https://www.latimes.com/archives/la-xpm-2006-sep-04-me-tharp4-story.html

127. *"Establishing the rift valley:* "Connect the Dots," by Marie Tharp.

128. *Today, acoustic instruments:* Wright, *ArcNews*.

128. *"Modern acoustic methods:* Ferrini, Vicki. 2019. "Seabed 2030: Mapping the Mysterious Deep." *Economist (World Ocean Initiative)*, December 9, 2019. https://ocean.economist.com /innovation/articles/seabed-2030-mapping-the-mysterious-deep

128. *What we've discovered so far:* Wright, Dawn. 2023. "To Protect the Oceans, We Must Map Them." *Mongabay*, August 10, 2023. https://news.mongabay.com/2023/08/to-protect-the-oceans-we-must-map-them-commentary

131. *"It is really something to see these crabs:* Wright, Dawn. 2021. "Women of the Deep" presentation, Explorers Club, April 2021.

132. *"We've got satellite altimetry:* From "Innovation of the Deep" ArcGIS StoryMaps story.

134. *"ROV operations are simpler:* NOAA. "What Is an ROV?" *NOAA Ocean Exploration Facts.* https://oceanexplorer.noaa.gov/facts/rov.html

135. *Autonomous Benthic Explorer:* Wright, "Women of the Deep" presentation (edited).

137. *Map the Gaps:* Map the Gaps website. https://www.mapthegaps.org

137. *"There's the expression cheap and deep:* Wright, NPR *Short Wave* interview.

137. *"We're working with various industries:* Wright, NPR *Short Wave* interview.

138. *"It's eye-opening:* Larson, Christina. 2024. "Largest Deep-Sea Coral Reef to Date Is Mapped by Scientists off the US Atlantic Coast." *Associated Press*, January 18, 2024.

138.   *"Has been right under our noses:* Larson, *Associated Press.*

138.   *"The generation of kids:* National Geographic, Education, "Ocean Exploration: Technology." https://education.nationalgeographic.org/resource/ocean-exploration

Chapter 8

139.   *"We saw tiny creatures:* Wright, Dawn. 2023. "Dawn Wright: Rolling in the Deep." Parley for the Oceans, April 20, 2023. https://parley.tv/journal/dawn-wright-challenger-deep

139.   *"I don't know why:* Wright, Dawn. 2023. "Dawn Wright: A Divine Abyss." *Biologos* podcast, May 18, 2023. https://biologos.org/podcast-episodes/dawn-wright-a-divine-abyss

140.   *"Our everyday world:* Black, Riley. 2023. *Deep Water,* p8. Chicago: University of Chicago Press.

140.   *"We live on an alien planet:* Black, p. 8.

140.   *"The thing that I love:* "Hollywood Filmmaker James Cameron on his Enduring Love of the Deep Ocean," by Stewart Campbell. *Boat International,* August 14, 2023. https://www.boatinternational.com/luxury-yacht-life/owners-experiences /james-cameron-on-his-love-of-the-ocean.

142.   *"Entire lineages of marine life:* "Two Thirds of Life in the Seabed Is Unknown to Science," by James Ashworth. Trustees of the Natural History Museum, London, February 4, 2022. https://www.nhm.ac.uk/discover/news/2022/february/two-thirds-life-seabed-unknown-science.html

144.   *"They are living with chemicals:* Wright, Dawn. 2023. "The Dive into Challenger Deep." Important, Not Important podcast, March 20, 2023.

145.   *"With new species consistently:* Alberts, Elizabeth Claire. 2023. "Seafloor Life Abounds around Hydrothermal Vents Hot Enough to Melt Lead." Mongabay, May 5, 2023. https://news.mongabay.com/2023/05/seafloor-life-abounds-around-hydrothermal-vents-hot-enough-to-melt-lead

145.   *"It's important to study:* Wright, Important, Not Important podcast.

145.   *"The triple threats:* Pratt, Monica. 2023. "Supporting the Science that Saves the Ocean." ArcUser, Spring 2023. https://www.esri.com/about/newsroom/arcuser/esriocean

145.   *UN Decade of Ocean Science:* The United Nations Ocean Decade. https://oceandecade.org

145.   *"The survival of our oceans:* Wright, Important, Not Important podcast.

146.   *"Trashing our only home:* Wright, Biologos podcast.

146.   *More than 400 million:* Bryce, Emma. 2023. "'We Can't Carry On': The Godfather of Microplastics on How to Stop Them." *The Guardian,* November 15, 2023.

146.   *Nearly a quarter of a million pieces:* "Plastic Particles in Bottled Water." *NIH Research Matters,* January 23, 2024. https://www.nih.gov/news-events-/nih-research-matters /plastic-particles-bottled-water#.

147.   *"The genus is Eurythenes:* Wright, Parley for the Oceans.

147.   *"Develop an international:* UN. 2023. "What You Need to Know about the Plastic Treaty Negotiations in Paris This Week." UN Environment Programme, May 29, 2023. https://www.unep.org/news-and-stories/story/what-you-need-know-about-plastic-treaty-negotiations-paris-week

147.   *"Once-in-a-planet opportunity:* Bryce, The Guardian.

148. *"If this goes wrong:* Baker, Aryn. 2021. "A Climate Solution Lies Deep Under the Ocean—But Accessing It Could Have Huge Environmental Costs." *Time*, September 10, 2021. https://time.com/6094560/deep-sea-mining-environmental-costs-benefits

149. *"We cannot, at this point:* Wright, Important, Not Important podcast.

150. *"By rushing in:* Baker, *Time*.

151. *"As though it's a Wild West:* Wright, Important, Not Important podcast.

151. *"Everything is connected:* Wright, Important, Not Important podcast.

151. *"Setting aside areas:* Wright, Dawn. 2023. "To Protect the Oceans, We Must Map Them." Mongabay, August 10, 2023. https://news.mongabay.com/2023/08/to-protect-the-oceans-we-must-map-them-commentary

152. *"Now it is our turn:* National Geographic. 2021. "We're a Young Explorer and a Scientist, and These Are Our Ocean Stories. What's Yours?" *National Geographic Education Blog*, September 2, 2021. https://blog.education.nationalgeographic.org/2021/09/02/were-a-young-explorer-and-a-scientist-and-these-are-our-ocean-stories-whats-yours

152. *"Everywhere that you are on the planet:* Wright, Parley for the Oceans.

152. *"The other big consequence:* Wright, Parley for the Oceans.

152. *It struck me while being:* Wright, Parley for the Oceans.

153. *"These changes are driving:* Sullivan, Kathy. 2020. "Why Exploration Could be the Key to Saving our Planet," with Dawn Wright. *Esri & The Science of Where* podcast, November 23, 2020.

153. *"People of color disproportionately:* Johnson, Ayana Elizabeth. 2020. "I'm a Black Climate Expert. Racism Derails Our Efforts to Save the Planet." *Washington Post*, June 3, 2020.

153. *"The quality of our natural world:* National Geographic Education Blog.

153. *"Mapping can ultimately:* National Geographic Education Blog.

153. *Some examples:* https://www.blackinmarinescience.org, bweems.org, https://oceandiscoveryleague.org, https://dx.doi.org/10.1029/2023AV000927, https://www.deepoceanobserving.org, www.wingsworldquest.org

155. *The following options:* mission-blue.org, https://mpatlas.org, livingatlas.arcgis.com, https://www.esri.com/en-us/about/science/maps-apps, https://www.esri.com/en-us/about/science/initiatives/ocean-science, algalita.org, albatrosscoalition.org, marine-conservation.org

156. *"Young people will take on:* National Geographic Education Blog.

157. *"[As we] human beings:* Thompson, Nainoa. 2023. "The Way of the Navigator." YouTube. https://www.youtube.com/watch?v=3TmvlM7B7dk

Image credits

Cover

Map. Challenger Deep diorama by Caladan Oceanic, GEBCO licensed under Creative
      Commons https://creativecommons.org/licenses/by/4.0/
Author photo. Photo courtesy of Dawn Wright.
  ix. Illustration by Victoria Roberts, Esri.

Chapter 1

2. Photo courtesy of Caladan Oceanic.
4. Illustration by Victoria Roberts, Esri, adapted from Gustavo Cardenas, Esri.
5. Photo courtesy of Caladan Oceanic.
6. Photo courtesy of Dawn Wright.
7. Photo courtesy of Dawn Wright.
8. Photo courtesy of Caladan Oceanic.
9. Photo courtesy of Dawn Wright.
10. Illustration by Victoria Roberts, Esri, adapted from Gustavo Cardenas, Esri.
11. *Top:* Photo courtesy of Caladan Oceanic.
11. *Bottom:* Photo courtesy of Caladan Oceanic.
12. Photo courtesy of Caladan Oceanic.
13. Photo courtesy of Caladan Oceanic. Cartography by Rochelle Wigley.
14. Photo courtesy of Caladan Oceanic.
15. Photo courtesy of Caladan Oceanic.
16. Photo courtesy of Caladan Oceanic.

Chapter 2

19. Photo courtesy of Dawn Wright.
21. Photo courtesy of Dawn Wright.
22. Roger C. Ambrose, The Magical Realm, RCA Designs. Photo courtesy of Dawn Wright.
23. Illustration by Robert Louis Stevenson, public domain, via Wikimedia Commons.
24. Photo courtesy of Dawn Wright.
25. Photo courtesy of Dawn Wright.
28. Photo courtesy of Dawn Wright.

30.   Photo courtesy of Dawn Wright.
33.   Photo from a video produced by Ari Daniel for Association for the Sciences of Limnol-
      ogy and Oceanography; American Geophysical Union. Illustration by Daniela Gamba.
35.   Photo courtesy of Dawn Wright.
36.   Photo courtesy of Dawn Wright.
39.   Photo courtesy of Esri.

Chapter 3

42.   *Top:* Cartography by John Nelson, Esri; image by Reto Stöckli, NASA Earth Observatory;
      Visible Earth. Tectonic plate data by Hugo Ahlenius, Peter Bird/Nordpil.
42.   *Bottom:* Cartography by John Nelson, Esri; image by Reto Stöckli, NASA Earth
      Observatory; Visible Earth. Tectonic plate data by Hugo Ahlenius, Peter Bird/Nordpil.
43.   Cartography by John Nelson, Esri; image by Reto Stöckli, NASA Earth Observatory;
      Visible Earth.
44.   *Top:* "The HMS *Challenger*," by William Frederick Mitchell, 1858. Public domain, via
      Wikimedia Commons.
44.   *Bottom:* Retrieved from NH 96801 US Navy bathyscaphe *Trieste* (1958–1963), Art collec-
      tion, US Naval History and Heritage Command website. Released by the US Navy Elec-
      tronics Laboratory, San Diego, California.
47.   Illustration by Victoria Roberts, Esri.
48.   *Left:* Cartography by John Nelson, Esri. Image by Visible Earth; NASA.
      Data by GEBCO Compilation Group (2021) GEBCO 2021 Grid (doi:10.5285/
      c6612cbe-50b3-0cff-e053-6c86abc09f8f.)
48.   *Right:* Cartography by John Nelson, Esri. Image by Visible Earth; NASA. Data
      by GEBCO Compilation Group (2021) GEBCO 2021 Grid (doi:10.5285/
      c6612cbe-50b3-0cff-e053-6c86abc09f8f.)
50.   Red sea serpent adapted from *Carta Marina* by Olaus Magnus, 1539. Public domain, via
      Wikimedia Commons.
53.   Reproduction from *National Geographic Magazine.* Photo by Leo Wehrli. CC BY-SA 4.0, via
      Wikimedia Commons.
54.   Photo by Mike Cole. CC BY 2.0, via Wikimedia Commons.
56.   Archival Photography by Steve Nicklas, NOS, NGS. Public domain, via Wikimedia
      Commons.

Chapter 4

59.   Photo by Kami Rita Sherpa. CC BY-SA 4.0, via Wikimedia Commons.
61.   Cartography by Victoria Roberts, Esri. Tectonic plate data by Hugo Ahlenius, Peter Bird/
      Nordpil.
62.   Photo by Reeve Jolliffe, Caladan Oceanic.
63.   Photo by Richard Varcoe, Caladan Oceanic.
64.   Photo by Richard Varcoe, Caladan Oceanic. CC BY-SA 4.0, via Wikimedia Commons.
65.   Photo by Verola Media, Caladan Oceanic.
66.   Photo by Special Project Six, Caladan Oceanic.

68. Photo by Victor Vescovo.
70. © Atlantic Productions/Discovery, from the Caladan Oceanic Five Deeps expedition.
73. Photo by Reeve Jolliffe, Caladan Oceanic.

Chapter 5

76. Cartography by Victoria Roberts, Esri. Trench data from National Geophysical Data Center and Esri.
77. Photo courtesy of Drew Stephens.
78. Photo courtesy of NASA.
80. Photo by Enrique Alvarez, Caladan Oceanic.
84. Photo by Nick Verola, Caladan Oceanic.
89. Photo courtesy of Dawn Wright.
91. Illustration by Victoria Roberts, Esri.
92. Photo courtesy of US Navy. Public domain, via Wikimedia Commons.
95. American Photo Archive / Alamy Stock Photo.
96. Photo courtesy of Caladan Oceanic.
97. Photo by Victor Vescovo. CC BY-SA 4.0, via Wikimedia Commons.
98. *Left:* Photo courtesy of Dawn Wright.
98. *Right:* Photo courtesy of Dawn Wright.
100. Photo courtesy of Dawn Wright.
101. *Top left:* Photo courtesy of Dawn Wright.
101. *Top right:* Photo courtesy of Dawn Wright.
101. *Bottom:* Photo courtesy of Dawn Wright.
102. *Top:* Photo courtesy of Victor Vescovo.
102. *Bottom left:* Photo courtesy of Victor Vescovo.
102. *Bottom right:* Photo courtesy of Caladan Oceanic.
104. Photo courtesy of Verola Media; Caladan Oceanic.
105. Photo courtesy of Dawn Wright.
106. Photo courtesy of Dawn Wright.
107. Photo courtesy of Dawn Wright.
108. Photo courtesy of Dawn Wright.

Chapter 6

111. Cartography by Cooper Thomas, Esri. Data by Peter Bird, Hugo Ahlenius/Nordpil.
113. Cartography by Cooper Thomas, Esri. Data by TeleGeography.
114. Photo courtesy of the DeepCCZ expedition; NOAA Ocean Exploration.
115. Graph by Victoria Roberts, Esri. Data by Frederikse et al. (2020). Adapted from NASA's Goddard Space Flight Center/PO.DAAC. https://www.nature.com/articles/s41586-020-2591-3
116. Cartography by Cooper Thomas, Esri. Data by IMF's World Seaborne Trade monitoring system (Cerdeiro, Komaromi, Liu, and Saeed, 2020).

118. *Source:* Cartography by Geoscience Australia, which is © Commonwealth of Australia and is provided under a Creative Commons Attribution 4.0 International License and is subject to the disclaimer of warranties in section 5 of that license. Data by Esri; Garmin International Inc.; US Central Intelligence Agency (The World Factbook); National Geographic Society | Esri, GEBCO, Garmin, NaturalVue.

Chapter 7

124. Illustration by Marie Tharp for *The Floors of the Oceans* (1959, fig 1).
125. Photo courtesy of Lamont-Doherty Earth Observatory and the estate of Marie Tharp.
127. Photo courtesy of the estate of Marie Tharp.
130. ©Woods Hole Oceanographic Institution, 1994.
131. ©Woods Hole Oceanographic Institution, 1991.
132. Illustration by Victoria Roberts, Esri, adapted from Gustavo Cardenas, Esri.
133. Photo courtesy of Monterey Bay Aquarium Research Institute (MBARI). © 2019 MBARI.
134. Photo courtesy of Caladan Oceanic.
135. ©Woods Hole Oceanographic Institution, 2002.

Chapter 8

141. Photo courtesy of Caladan Oceanic.
143. Image courtesy of the NOAA Office of Ocean Exploration and Research, 2017, American Samoa.
147. Copyright 2017–2019. Adobe Stock. All rights reserved.
149. Image courtesy of the NOAA Office of Ocean Exploration and Research, 2015, Hohonu Moana.

About Esri Press

Esri Press is an American book publisher and part of Esri, the global leader in geographic information system (GIS) software, location intelligence, and mapping. Since 1969, Esri has supported customers with geographic science and geospatial analytics, what we call The Science of Where®. We take a geographic approach to problem-solving, brought to life by modern GIS technology, and are committed to using science and technology to build a sustainable world.

At Esri Press, our mission is to inform, inspire, and teach professionals, students, educators, and the public about GIS by developing print and digital publications. Our goal is to increase the adoption of ArcGIS and to support the vision and brand of Esri. We strive to be the leader in publishing great GIS books, and we are dedicated to improving the work and lives of our global community of users, authors, and colleagues.

Acquisitions

Stacy Krieg
Claudia Naber
Alycia Tornetta
Craig Carpenter
Jenefer Shute

Editorial

Carolyn Schatz
Mark Henry
David Oberman

Production

Monica McGregor
Victoria Roberts

Sales & Marketing

Eric Kettunen
Sasha Gallardo
Beth Bauler

Contributors

Christian Harder
Matt Artz

Business

Catherine Ortiz
Jon Carter
Jason Childs

Related titles

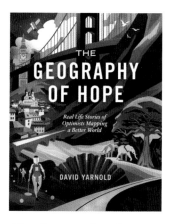

The Geography of Hope

David Yarnold

9781589487413

The Power of Where

Jack Dangermond, Esri,
and the GIS Community

9781589486065

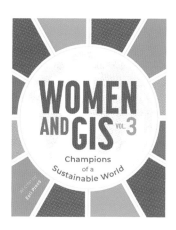

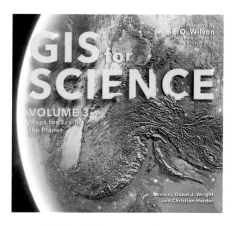

Women and GIS, volume 3

Esri Press

9781589486379

GIS for Science, volume 3

Dawn J. Wright and Christian Harder
(editors)

9781589486713

For more information about Esri Press books and resources, or
to sign up for our newsletter, visit

esripress.com.

.